NOTES ON
PRACTICAL MECHANICAL
DRAWING

WRITTEN FOR THE USE OF STUDENTS
IN ENGINEERING COURSES

BY

VICTOR T. WILSON, M. E.

PROFESSOR of DRAWING AND DESIGN at the
MICHIGAN AGRICULTURAL COLLEGE,
EAST LANSING, MICHIGAN

AND

CARLOS L. McMASTER, C. E.

Formerly ASSOCIATE io GENERAL ENGINEERING DRAWING
at the UNIVERSITY OF ILLINOIS
URBANA, ILLINOIS

FOURTH EDITION
REVISED AND ENLARGED

British Library Cataloguing-in-Publication Data
A catalogue record for this book is available from the
British Library

Technical Drawing and Drafting

Technical drawing, also known as 'drafting' or 'draughting', is the act and discipline of composing plans that visually communicate how something functions or is to be constructed.

It is essential for communicating ideas in industry, architecture and engineering. The need for precise communication in the preparation of a functional document distinguishes technical drawing from the expressive drawing of the visual arts. Whereas artistic drawings are subjectively interpreted, with multiply determined meanings, technical drawings generally have only one intended meaning. To make the drawings easier to understand, practitioners use familiar symbols, perspectives, units of measurement, notation systems, visual styles, and page layout. Together, such conventions constitute a visual language, and help to ensure that the drawing is unambiguous and relatively easy to understand.

There are many methods of constructing a technical drawing, and most simple among them is a sketch. A sketch is a quickly executed, freehand drawing that is not intended as a finished work. In general, sketching is a quick way to record an idea for later use, and architects sketches in particular (in a very similar manner to fine artists) serve as a way to try out different ideas and establish a composition before undertaking more finished work. Architects drawings can also be used to convince clients of the merits of a design, to enable a building constructer to use them, and as a record of completed work. In a similar manner to engineering (and all other technical drawings), there is a set of conventions (i.e particular views, measurements, scales, and cross-referencing) that are utilised.

As opposed to free-sketching, technical drawings usually utilise various manuals and instruments. The basic drafting procedure is to place a piece of paper (or other material) on a smooth surface with right-angle corners and straight sides – typically a drawing board. A sliding straightedge known as a 'T-square' is then placed on one of the sides, allowing it to be slid across the side of the table, and over the surface of the paper. Parallel lines can be drawn simply by moving the T-square and running a pencil along the edge, as well as holding devices such as set squares or triangles. Other tools can be used to draw curves and circles, and primary among these are the compasses, used for drawing simple arcs and circles. Drafting templates are also utilised in cases where the drafter has to create recurring objects in a drawing – a massive time-saving development.

This basic drafting system requires an accurate table and constant attention to the positioning of the tools. A common error is to allow the triangles to push the top of the T-square down slightly, thereby throwing off all the angles. Even tasks as simple as drawing two angled lines meeting at a point require a number of moves of the T-square and triangles, and in general drafting this can be a time consuming process. In addition to the mastery of the mechanics of drawing lines, arcs, circles (and text) onto a piece of paper – the drafting effort requires a thorough understanding of geometry, trigonometry and spatial comprehension. In all cases, it demands precision and accuracy, and attention to detail.

Conventionally, drawings were made in ink on paper or a similar material, and any copies required had to be laboriously made by hand. The twentieth century saw a shift to drawing on tracing paper, so that mechanical copies could be run off efficiently. This was a substantial development in the drafting process – only eclipsed in the twenty-first century with 'computer-aided-drawing' systems (CAD). Although classical draftsmen and women are still in high demand, the mechanics of the drafting task have largely been automated and accelerated through the use of such systems. The development of the computer had a major impact on the methods used to design and create technical drawings, making manual drawing almost obsolete, and opening up new possibilities of form using organic shapes and complex geometry.

Today, there are two types of computer-aided design systems used for the production of technical drawings; two dimensions ('2D') and three dimensions ('3D'). 2D CAD systems such as AutoCAD or MicroStation have largely replaced the paper drawing discipline. Lines, circles, arcs and curves are all created within the software. It is down to the technical drawing skill of the user to produce the drawing – though this method does allow for the making of numerous revisions, and modifications of original designs. 3D CAD systems such as Autodesk Inventor or SolidWorks first produce the geometry of the part, and the technical drawing comes from user defined views of the part. This means there is little scope for error once the parameters have been set.

Buildings, Aircraft, ships and cars are now all modelled, assembled and checked in 3D before technical drawings are released for manufacture.

Technical drawing is a skill that is essential for so many industries and endeavours, allowing complex ideas and designs to become reality. It is hoped that the current reader enjoys this book on the subject.

PREFACE

This book is a collection of notes intended to furnish the basis for a course in elementry mechanical drawing, so arranged, it is thought, that the teacher may have the widest latitude in his choice of sequence of subjects.

Since its first edition, two years ago, the book has been rearranged with this particular point in view. It has been thoroughly revised and also enlarged by the addition of more explanatory matter and illustrations in orthographic projection, by a chapter upon isometric and oblique drawing, and by a number of exercises in working drawings from sketches. The usual geometrical drawing has been reduced to a minimum; it has been included, not for its value as exercise in drawing, but for the knowledge conveyed upon constructive processes useful in the science.

Latest practice in teaching drawing shows the influence of utilitarianism. The aim of a mechanical drawing is to record useful facts; useful facts, therefore, are used as exercises from the beginning of the study. The theoretical or geometrical science forms but a very small part of the knowledge required in the subject and, where courses in drawing are necessarily short, the maximum of practical information is aimed at. The course here outlined uses working drawings as exercises almost from the very beginning.

Of course, the actual subject, from which to draw, is most to be desired, but it is not always available or practicable, particularly in the beginning and with large classes, therefore numerous working drawing sketches are presented as problems. Again, nothing can take the place of the personal guidance of the teacher in inculcating good methods of work, but large classes make the individual direction by the teacher difficult, hence, minute directions are given upon the care and use of tools and methods of working to aid the teacher and student. And it may be stated, from their experience, it is the conviction of the authors, that, where good methods and system in working are insisted upon from the beginning, the quality of the the product is greatly improved. This theory is opposed to the one that practice only, together with quantity, makes perfect.

Of course it is expecting a great deal of the student that he grasp all the practical points given in a short course, but, here again, experience has verified the conviction that it is possible and especially where the exercises are made to be of interest, for where interest is aroused individual initiative will do the rest.

Quotations have here and there been made from standard works on drawing, and acknowledgment has been given in each case.

V. T. WILSON.

C. L. McMASTER.

TABLE OF CONTENTS.

CHAPTER V.

WORKING DRAWINGS.

CHAPTER VI.

GEOMETRICAL DRAWING.

CHAPTER VII.

MACHINE SKETCHING.

APPENDIX.

INDEX

NOTES ON
PRACTICAL MECHANICAL DRAWING.

CHAPTER I.
LETTERING.*

1. The lettering, which the draftsman, in practice, uses most, is a rapidly executed statement, on a drawing, in what is known as an off-hand style. It is a very simple letter which he learns, with practice, to do in ink without any preliminary pencil layout, beyond the limiting lines, to show the height of the letters.

Before the beginner, however, can hope to attain a proficiency, in even the free lettering mentioned, he should study carefully letter forms as they have been gradually developed through the centuries, to what are called standard proportions.

2. Good lettering is not mechanical, but is good design. The straight edge, compass or other tools, have no place in the drawing of letters, beyond the making of the limiting lines just mentioned. Good design requires: (*a.*) *Simplicity of style*, instanced in the advertisements confronting us so commonly every day. (*b.*) *Uniformity of*

---

*The matter upon lettering is extracted from "Free-hand Lettering," by Victor T. Wilson, by kind permission of the publishers, Messrs. John Wiley & Sons.

effect, as the units in any design are distributed throughout the area to be covered. The letters, that is, should
appear to be of the same height, the same general size,
and the spacing should also appear to be uniform. These
things can only be properly attained through judgment
and taste combined with accuracy of eye in the detection
of small differences.

3. Letters actually vary in width, because those which do
not fill their rectangle of space, as the H does, look smaller
if they are made of the same width as the H. They must
be made slightly wider than the normal letter; for
example, the letter A must be spread at the base, because
it only occupies half the rectangle of space allotted to it;
likewise the B, C, D, etc., must be widened, each to a different degree. The exceptions to this are the L and the
F, which are made narrower than the normal letter.

4. Letters actually vary in height, because, where a
letter touches its upper and lower limits only by tangency,
it would look shorter than the H or N, if made tangent to
them; it must be made to slightly exceed both limiting
lines, as the C, G, and O. Letters such as A, V, etc.,
should also exceed the limits, if their angles are made
sharp; to overcome this they are often somewhat blunted

5. Letters are modified to produce an effect of stability;
that is, those letters that have upper and lower parts, dis-
tinctly separated, appear more stable, and of good form,
if the lower section is made larger than the upper; for
example, the lower lobe of the B, the spread of the arms

of the X, the K, the lower horizontal stroke of the E and the Z; the lower curve of the S, also, is made larger across and higher than the upper.

6. Letters are further varied in their several variations; that is, when combined into words, slight modifications can be introduced, here and there, to advantage; for example, an L, just preceeding an A, can be made narrower than if it were followed by an H or were at the end of a word or a line. To mechanically figure out all these modifications simply spoils the spontaniety of design; they must be the result of judgment applied in each particular case. Hence, the student should not regard the modifications shown on the plates as having any value, beyond suggestions to aid in the formation of correct perceptions.

7. A knowledge of free-hand drawing is essential to facility in lettering because the eye is then trained to see form and to judge of effects; moreover lettering should be developed much as a free-hand drawing is developed, by first getting a broad, simple effect and proceeding to the details gradually in the order of their importance.

8. Figure 1 shows the upper case Roman letter, in a standard form. *Upper and lower case* are terms used by printers for capital and small letters, so named, because the type representing them are placed, respectively, in the upper and lower part of the type case.

The Roman letter has no really standard form, in which exact proportioning is attainable. The ancestors of this letter had a very different form from that we now find

FIGURE No. 1.

MODERN ROMAN.

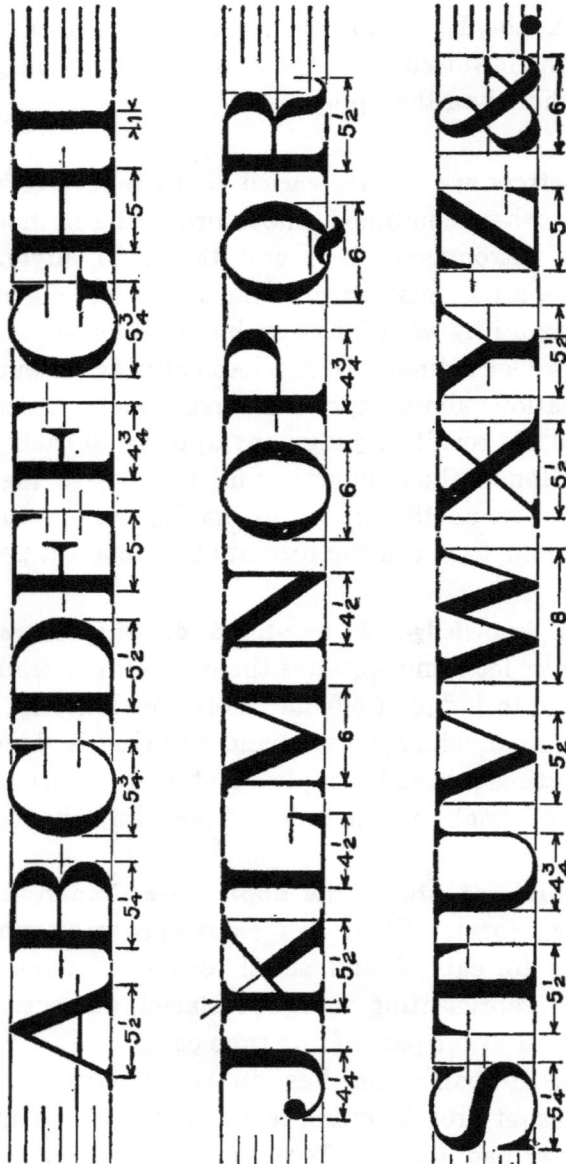

in the printer's type, or in modern good examples. It has
been modified and changed by different authorities, we
cannot point to any one illustration of a perfectly correct
Roman type, but to many, varying slightly in some cases,
quite radically in others. This letter is a refinement of an
imitation of the strokes of the quill used by the early
scribes. It is nearly square, in fact was square, in its
early forms.

Referring to the figure, the heavy stems are made a
normal width of one unit, the height of the letter being
divided into six parts called units. The numerals at the
bottom of each letter also stand for units. If the body of
a letter varies in thickness, as the B, C, G, etc., the maximum width, at the middle, is slightly greater than one
unit, the S and U being exceptions.

The large spurs on the E, F, L, T and Z do not join
the body of the letters like the serifs, by tangent curves;
they meet the horizontal strokes abruptly.

The mid-horizontal strokes of the B, E, F, H and R
are put slightly above the center of the space, to lend an
effect of stability, the P is an exception to this.

The inner and outer edges of the curved parts of
letters, as B, C, O, P, and the upper part of the letter R,
are formed by arcs of regular curves with vertical and
horizontal axes; the inner ones approach the outer
tangentially. The vertical axes of the outer curves are
slightly larger than their horizontal ones, except the U.

The difficulties of drawing the S, common to beginners,
may be materially lessened by using an O, of the same
proportions, as a basis in sketching.

FIGURE No. 2.

ROMAN AND GOTHIC SMALL.

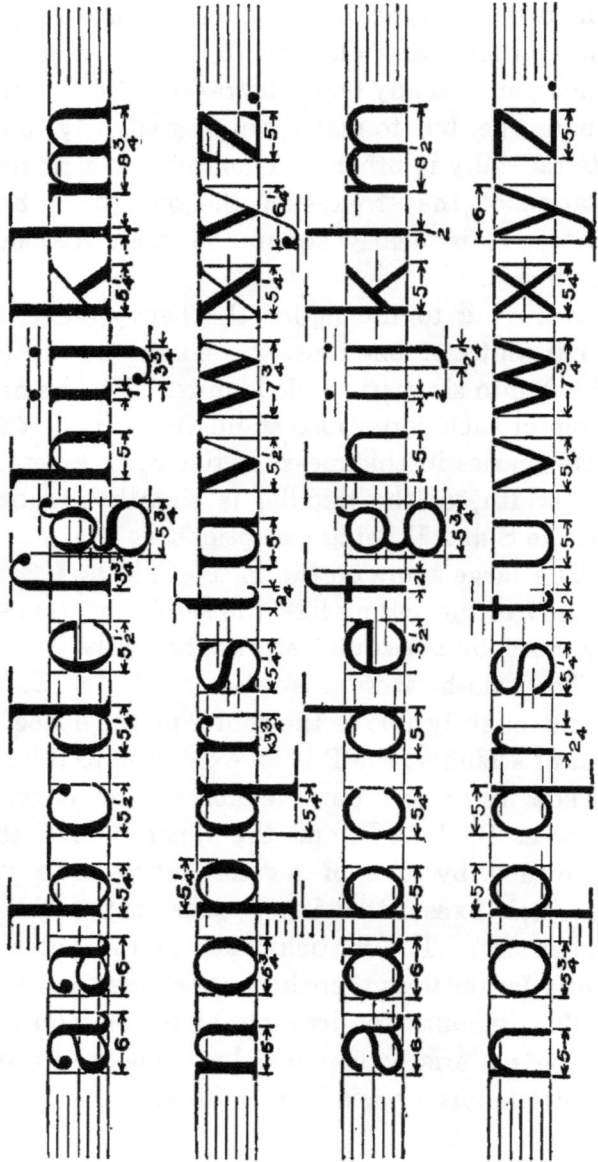

9. The small letters are shown in Fig. 2, to harmonize with the capitals. These may be divided into three classes, *ascending, descending and short letters.* The ascending, except the 't,' have a height equal to the capitals, and the descending are the same in total length. The height of the short letters, relative to the others, is not fixed, but they generally vary between about one-half and two-thirds the height of the capitals. In the figure they are six-tenths.

The width and height of the small letters are related to each other, in the same manner as the corresponding dimensions of the capitals. The height of the short letters is divided into six parts, each a unit for both the width of the letter and the weight of body. The same peculiarities, as to variations, which occur in the capitals, also occur in the small letters.

FIGURE No. 3

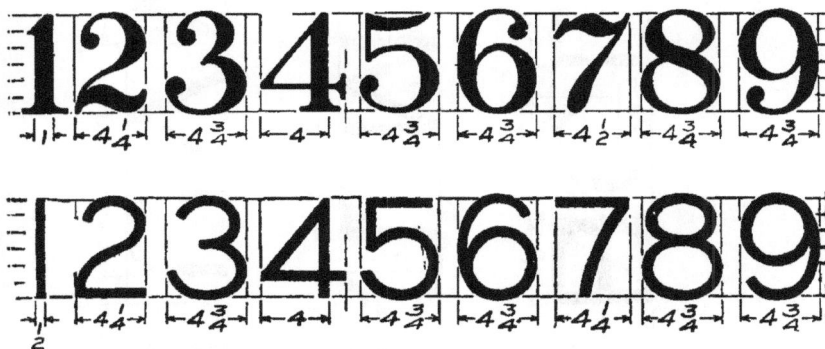

10. The Roman numerals are shown in Fig. 3. They can be made the height of the capitals or slightly shorter, according to taste. They have the same peculiarities, as to variation in height, width and weight of body, that the letters do.

FIGURE No. 4.

MODERN GOTHIC.

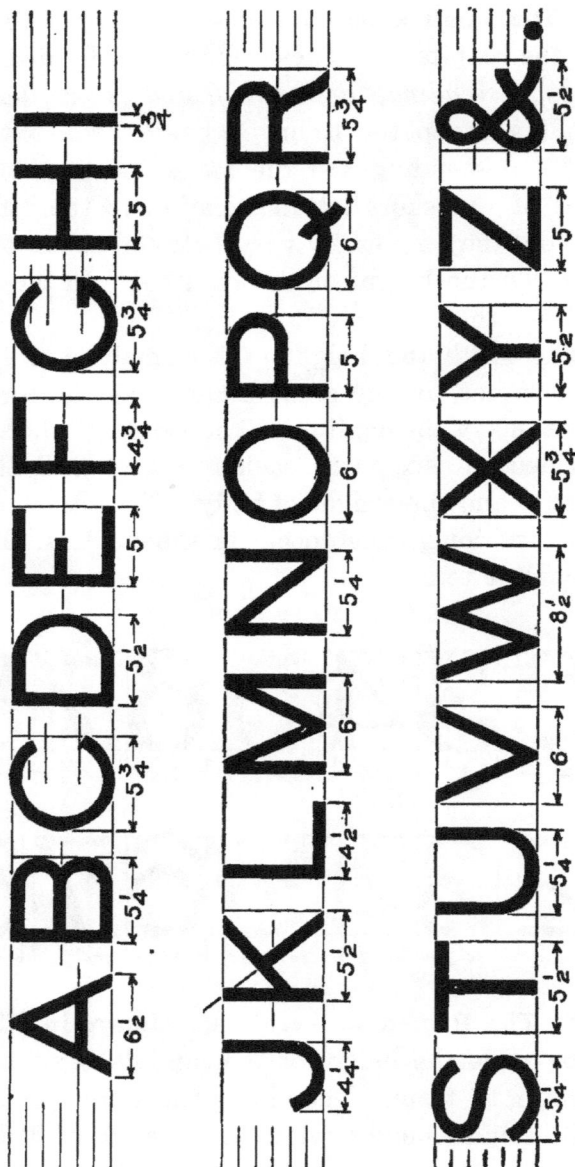

11. The Gothic or uniform body letter, as shown in Fig. 4, is one of most common use by draftsmen, either as a heavy body or a light one made by a single stroke of the pen. For a letter to appear as heavy as the Roman, the Gothic is made slightly thinner in the stems. Note that the ends of all letters are cut off perpendicularly to the outside edge of the body. It has the same variations as noted in the Roman.

12. Off-hand lettering is shown in Fig. 5, in the simplest and probably the most common style, the Gothic. It should be made directly with the pen, the limiting lines only being ruled; this last, the beginner should never fail to do.

It has been found that a certain system in strokes produces the best results and such an analysis is given in the upper rows, showing several ways of doing it, any one of which is good. It requires a great deal of skill to handle off-hand lettering satisfactorily, and the beginner should not be discouraged at the large amount of practice work necessary to attain it.

Cultivate a steady, uniform stroke, as far as possible, towards the person, as a basis for these letters. It is impracticable to depend upon patching unsatisfactory lines.

Rows 4 and 5, of Fig. 5, are the same letter as that in the first three rows, and is of a standard proportion and quite common size. Rows 6 and 7 are the same letter inclined. The inclination should be about 20° from the vertical; it can be somewhat greater, but it should not be less, else it is apt to look like a poor attempt at the vertical

FIGURE No. 5.

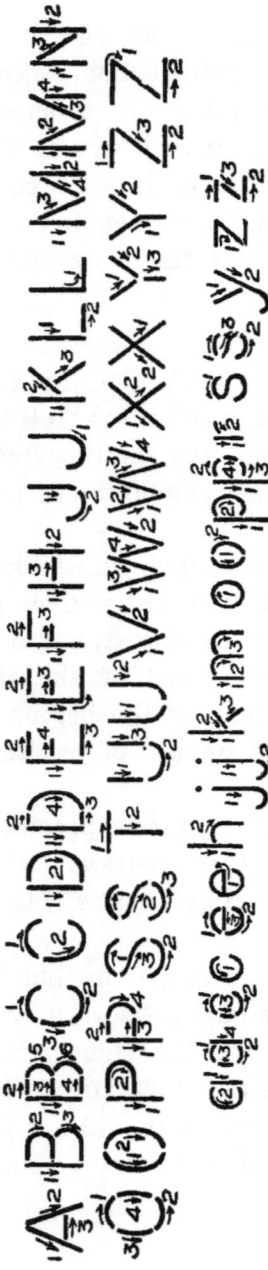

ABCDEHIJKLMNOPRSTUVWXYZ

abcdefghijklmnopqrstuvwxyz.

ABCDEHIJKLMNOPRSTUVWXYZ

abcdefghijklmnopqrstuvwxyz.

abcdefghijklmnopqrstuvwxyz

style. An important point to note, in the inclined style, is that V and W, etc., are so inclined that the bisector of the angle of the sides has the direction of the main stems of the other letters. Also, note that the O forms, if accurately made, are inscribable in a parallelogram, whose inclined sides have the direction of the main stems of the other letters; that is, the axes of the oval forms do not have exactly this direction.

The kind of pen used for this letter will determine the weight of the strokes. At first it will be found difficult to make a continuous straight stroke of uniform weight; to aid in doing this; *first, hold the pen so that the plane of the pen axis, and the line to be made, are perpendicular to each other, then touch the paper, pressing the nibs of the pen apart to the proper width before starting the stroke; after starting, continue motion uninterruptedly until the end, and lift the pen just an instant after stopping motion, else the line will taper to a fine point. If a lump accumulates upon either end of the line, it can be overcome by carrying less ink in the pen, combined with a briefer hesitation at beginning and ending. Whole arm motion is found helpful.*

Fig. 6 shows some other styles of off-hand lettering. Rows 1 and 2 are based upon the Gothic and have its characteristics, mainly, while rows 3 and 4 show a free style appropriate to architectural drawings. The proportions of letters used depend upon the space allowable for them; however, a broad letter finds more favor than a narrow one. It is certain that increasing the breadth of a letter increases its legibility more than a corresponding and proportional increase in height. The expanded form is shown in several places later on.

FIGURE No. 6.

ABGDEHIJKLMNOPRSTUVWXYZ

abcdefghijklmnopqrstuvwxyz.

*ABCDEFGHIJKLMNOPQRSTUVWXYZ

abcdefghijklmnopqrstuvwxyz.&

	Wr't I:	
Connect. Rod	Wr't I:	1
,, Strap	,,	1
,, Brass	Brass	2
,, Wedge	C.I.	2
,, Bolt	Mch·S·	2
Nut for ,,	,,	6
Adjust· Screw	,,	1
Brass for Cr·E·	Brass	1
Plate ,, ,,	,,	1
Brass for Cyl·E·	,,	1
Plate ,, ,,	,,	1
Mch·Scr· ,, ,,	Mch·S·	4
Wedge ,, ,,	C.I.	1

BILL OF MATERIALS.

No	NAME	DIMENSIONS	B·M·	Total.
2	Plumb Posts	12×12×14	168	336
2	Batter Posts	12×12×16	192	384
1	Sill	12×12×14	168	168
1	Cap	12×12×12	144	144
3	Bolsters	9×12×3	12	36

* From 'Plain Lettering' by Henry S. Jacoby.

13. The desirable size for an off-hand letter depends upon circumstances; if it is made, for example, from one to one and one-half times as wide as it is high, the short letters may be one-sixteenth of an inch in height, as it is made narrower the height must be increased, for equal legibility, in greater proportion. The usual sizes may be said to vary between one-sixteenth and one-eighth of an inch.

14. The desirable pen to use, may be a *ball pointed pen*, if fine, or a Gillott's 303; the latter should be worn down a little to produce the most satisfactory results. The *falcon pen* is also a good one; any pen of medium stiffness, will do which will make the desired weight of stroke without patching. *In wiping the pen*, be careful not to press the nibs too far apart against the wiping cloth, they are apt to take a set and spoil the spring in the pen. *In filling the pen*, use the quill attached to the cork, in preference to dipping the pen in the bottle, as the handle is apt to get ink on it. *Clean the pen* every time it has to be filled and *keep the ink bottle closed, tight*, except when filling pen.

15. The old Roman letter is shown in Fig. 7, rows 1 to 5. It is an attempt to preserve as much as possible of the early type of the Roman as is consistent with modern requirements in letters. It is a beautiful form and is increasing in popularity; it is a much freer letter than the conventional Roman, and, being free, it is not found in such a uniform style, there are more variations indulged in. The letter here shown is a very careful compilation from the best authorities.

FIGURE No. 7.

ABCDEFGHIJK
LMNOPQRSTV
WX
YZ.

abcdefhiklmnorst
gjpquvwxyz.

ABCDEJKLNRSVZ.
abcdf g jknprtuvy23456

16. **The Roman-Gothic letter** is shown in rows 6 and 7. It is a very popular form of letter today, being a combination of two styles, as its name implies. It probably is preferred, because of its heavy face and consequent ease of construction.

17. **Titles to drawings** are placed, commonly, in the right-hand lower corner and consist of (*a*) the designation for the drawing, (*b*) the firm for whom it is made, (*c*) the scale, (*d*) the date, and frequently a set of items such as 'checked by,' 'corrected,' 'traced by,' etc.

In designating titles to drawings, the fundamental requisite is *appropriateness;* simplicity in style is an adjunct to this for working drawings, and the simplest letter is the Gothic.

Titles are generally made wholly in capitals, although there is no rule to control this. If small letters are used, it is, generally, where minor matter has to be compressed into small space.

It is well to lay out the whole title in a design with a very simple contour shape, and give a suggestion only of the weight of the different lines. Find the middle or axial line, of the title and sketch in on both sides of it. Fig. 8 shows a method of laying out a simple title.

Make minor connectives small and, if short, preferably use an expanded letter. The title should look equally well if all connectives, as 'made by,' 'for,' etc., are left out.

Use only a few styles of letters in a title, not more than three, preferably one, and do not depend upon

FIGURE No. 8.

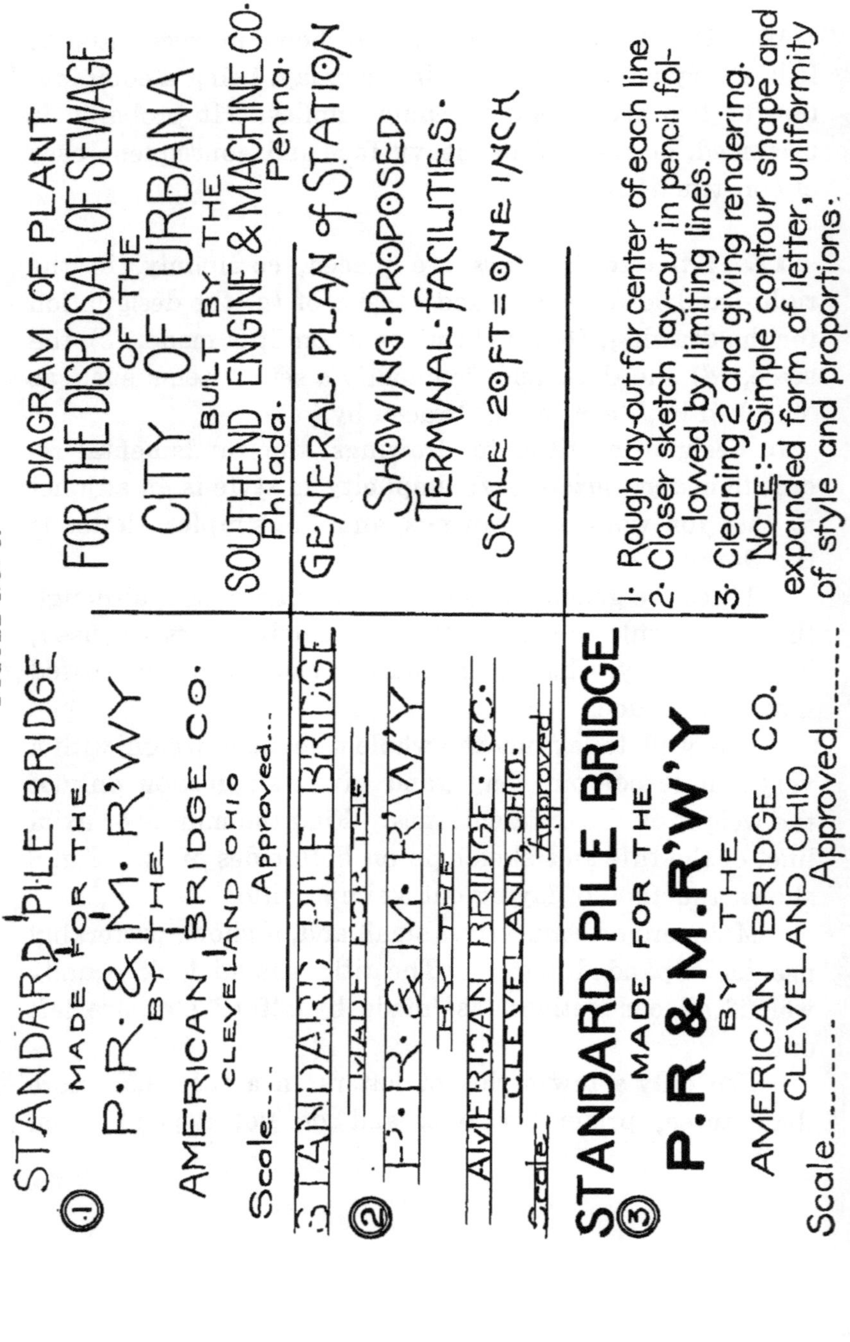

DIAGRAM OF PLANT
FOR THE DISPOSAL OF SEWAGE
OF THE
CITY OF URBANA
BUILT BY THE
SOUTHEND ENGINE & MACHINE CO.
Philada.   Penna.

GENERAL · PLAN · of STATION
SHOWING · PROPOSED
TERMINAL · FACILITIES ·
SCALE 20 FT = ONE INCH

1. Rough lay-out for center of each line
2. Closer sketch lay-out in pencil followed by limiting lines.
3. Cleaning 2 and giving rendering.
   NOTE:— Simple contour shape and expanded form of letter, uniformity of style and proportions:

① STANDARD PILE BRIDGE
MADE FOR THE
P·R·&·M·R'WY
BY THE
AMERICAN BRIDGE CO.
CLEVELAND OHIO
Scale___   Approved___

② STANDARD PILE BRIDGE
MADE FOR THE
P·R·&·M·R'WY
BY THE
AMERICAN BRIDGE · CO·
CLEVELAND, OHIO.
Approved___
Scale

③ STANDARD PILE BRIDGE
MADE FOR THE
P.R.&M.R'W'Y
BY THE
AMERICAN BRIDGE CO.
CLEVELAND OHIO
Scale___   Approved___

FIGURE No. 9.

PLAN OF

PROPOSED EXTENSION

of the

WATER SUPPLY SYSTEM

of the

CITY OF LAWRENCE

·Showing the connections with existing mains.

Prepored under the direction of

Chas. H.Hill City Eng'r.

1907

Scale ~ Feet

100 50 0 100 200 300 400

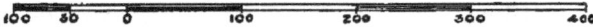

LOCATION SURVEY

PLAN AND PROFILE

FOR THE

MEMPHIS BRANCH

OF THE

SOUTHERN RAILWAY

Prepared by the

Engineering Department

Nashville – Tenn.

June 1908

Scale of Feet

100 50 0 200 400 600 800

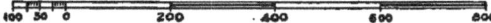

variety of treatment but upon harmony, neatness, compactness and legibility.

The two titles of Fig. 9 illustrate the style common in civil engineering and map drawing. More painstaking work is characteristic of civil engineering practice in lettering than is found in other professions. The second title is of the simplest style. The first is more difficult, being the Roman letter throughout.

18. Combinations of Roman and stump letters, either verticle or inclined, are found in the majority of titles on maps of all kinds, including municipal maps. The difficulty of construction limits its field to the higher class of finished drawings. The letter may also be executed very acceptably off-hand. Such titles are very common in engineering practice.

The stump letter is shown in the last two rows on Fig. 5, the capitals, for which, are the inclined Roman.

19. A bill of materials is often attached to drawings. It may be in a corner of the sheet, and usually the lower right-hand corner, above the title or it may be upon a separate sheet or sheets, if it is very extended. It is made in off-hand lettering, neatly executed. Its contents will be further discussed in Chapter IV. The lower part of Fig. 6 gives examples of such tabulation.

CHAPTER II.

ORTHOGRAPHIC PROJECTION.

20. Drawing is the art or science of recording a person's impressions about things by a more or less accurate suggestion of form. It is a technical subject of great value to the engineer and architect, the graphical language with which they work and by which they convey their ideas to those who construct the things they lay out. It is the language, moreover, understood by all nations, having for its purpose the complete representation of any object or structure to be erected.

All drawings may be divided into two general classes. (*a.*) The drawings of objects as viewed at a finite distance. (*b.*) The drawings of objects as viewed at an infinite distance. The first of these is called *perspective*. The point of view at a finite distance is called the *center of projection*. It is as if the eye were at the point and the drawing of the object was made upon a transparent plane placed between the latter and the center of projection, that is, projected upon it from this center by lines from the center passing through all points of the object. It is a kind of drawing that is found in pictures. In the second class the center of projection or the eye is theoretically moved to an infinite distance, that is, the projecting lines, from the object to the plane, become parallel. This, in a certain form of drawing, is what is called *orthographic projection*.

Things are constructed and manufactured from drawings made according to the principles of the second kind mentioned, or orthographic projection. They may be made free-hand, that is by sketches, or they may be made by careful mechanical drawings.

A mechanical drawing, used for the purpose of construction, then, consists of one or more views, made according to the principles of orthographic projection; in addition to which the sizes of parts are clearly set forth by dimensions, notes or other symbols that are required to accurately construct the same.

No matter how simple is the subject to be constructed, an accurate and comprehensive and unmistakable drawing should be made of it. The test of a good working drawing lies in the fact that the workman can make nothing out of the facts contained thereon than what was intended by the draftsman. The entire meaning should be clear beyond the shadow of a doubt. To choose the number of views that this may be attained, to put on the dimensions which the workman will need in making the subject, is the problem of the draftsman. The needs of the workman should be constantly in his mind.

21. Orthographic projection is the science of representing forms by projecting them upon two or more planes by means of projecting lines respectively perpendicular to these planes.

The center of projection, from which the projecting lines eminate, is infinitely distant in a direction perpendicular to the planes of projection, and, since they pass through a common point at infinity, they are parallel.

The 'coordinate planes of projection,' is a term used to designate the three fundamental planes in common use, namely, a vertical plane (V), a horizontal plane (H), and

FIGURE NO. 10.

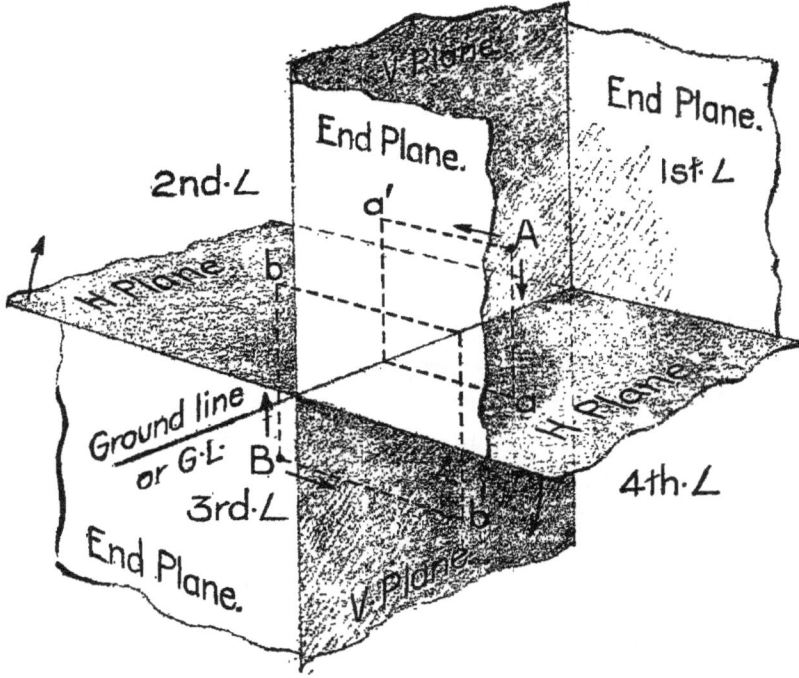

a plane perpendicular to these two, known as the end plane, or profile plane (E or P). The vertical and the horizontal planes intersect each other in a line known as a 'ground line' (G.L.) and each is indefinite in extent.

Fig. 10 shows a picture as in perspective, of these

several planes as they would appear if made of transparent material.

22. The V and H planes form with each other four dihedral angles, but, for purposes of representation on a drawing, it is necessary to conceive of them as being folded into coincidence with each other. When so folding, they revolve about the G.L. in the direction of the arrows shown, the H plane being revolved into coincidence with the V plane, or the latter revolved into coincidence with the H plane so that the 2nd and 4th angles, as designated in Fig. 10, close up to 0°, while the 1st and 3rd, unfold to 180°. The revolution of the end or profile plane will be discussed a little later.

*23. **A is a point in space,** as shown in Fig. 10 in the first dihedral angle. Its projection on the V plane is the foot of the perpendicular to the V plane Aa'. Likewise its projection on the H plane is the foot of the perpendicular to the H plane Aa.

If perpendiculars are dropped from a' and a respectively to the G.L., they will intersect the G.L. in a point, and the four lines together make a rectangle.

24. The distance of a' from the G.L. shows the distance of the point from the H plane and the distance of a from the G.L. shows the distance of the point from the V plane. When the V and H planes are folded into coincidence, as in Fig. 11, the perpendiculars, let fall to the G.L. from a'

---

*The notation, which will be used, is as follows: — a', b', c', etc., stand for the vertical projections of points, and a, b, c, etc., stand for the horizontal projections of the same points.

and *a*, become one and the same line perpendicular to the
G.L., and it is in this form that orthographic projection
deals with points, lines, etc.

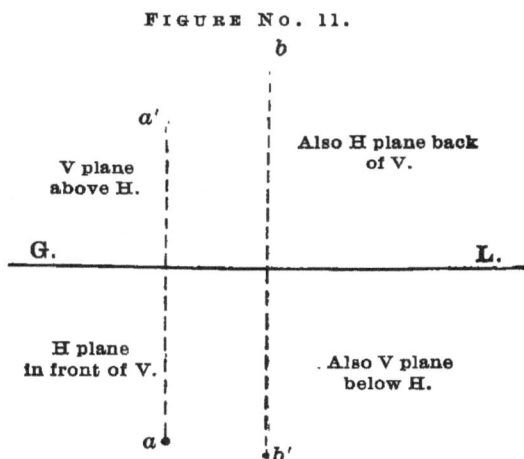

25. **The projections of a point** in the 1st angle, 2 inches
from V and 3 inches from H, in V projection, would be
placed 3 inches above the G.L., and in H projection would
be put on a perpendicular to the G.L. through the V projection, a distance of 2 inches below the G.L.

If it were required to show the projections of a point
which was in the 3rd angle, 1½ inches from V, and 4 inches
from H, the V projection would be placed 4 inches below
the G.L., and on the same perpendicular to the G.L.,
through it, would be placed the H projection 1½ inches
above the G.L.

If a point is in one of the co-ordinate planes, its corresponding projection is in the G.L., that is, if a point is in
the V plane, its H projection will be at the intersection
with the G.L. of a perpendicular through its V projection.
If a point is in both co-ordinate planes, it must lie at their

intersection, i. e., be in the G.L., and both projections coincide with each other, and with the point itself.

Further discussion of the principles will be confined to forms in the 3rd angle exclusively as this is more common in practice.

26. Lines are projected by projecting points on the line; if straight lines, the projections of two points will locate the projections of the line, and designate the position of the line in space with respect to the co-ordinate planes.

FIGURE NO. 12.

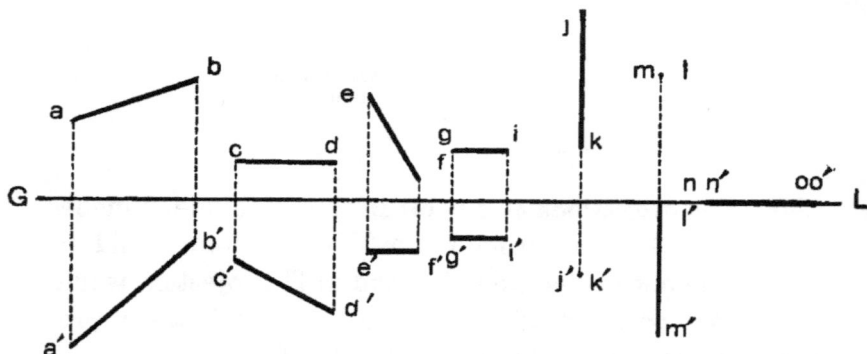

Fig. 12 shows the several projections of a limited line and also embodies all the different positions which a line can have with respect to the co-ordinate planes in the third angle. A B is oblique to both V and H, since the distances of A from both V and H are different from those of B.   CD is parallel to V and oblique to H because the distances of C and D from V are equal; JK is perpendicular to V since the line connecting *j* and *k* is perpendicular to the G.L., and coincides with the projecting line of both J and K on V. NO lies in both H and V, i. e. in the G.L., and both projections of N and O, respectively, coincide with each other. This group of projections may be called, appropriately, the alphabet of the straight line in the 3rd angle.

If a line is parallel to a coordinate plane its projection on that plane will show, (a), the true length of the line, and (b), by its angle with the G.L., the angle the line makes with the other or corresponding coordinate plane.

27. The projections of a plane figure are obtained by projecting separately points in the perimeter of the figure and connecting the projections by lines. If the figure is composed of a curve or curves, the projections of a convenient number of points of the curve are connected by smooth curves. If the figure is composed of straight lines, only the vertices of the figure, or meeting points of the edges, need be projected.

FIGURE No. 13.

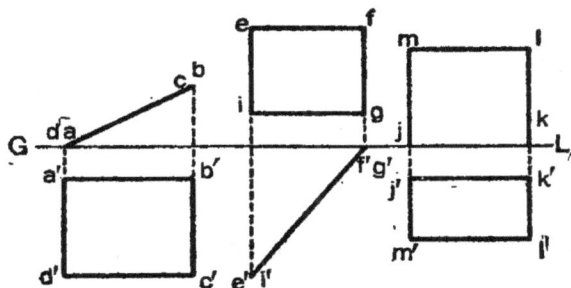

Fig. 13 illustrates the projections of a plane rectangular figure in different positions in the 3rd angle, (a) when it is perpendicular to H and oblique to V, with two edges, AB and DC, parallel to H, (b) when it is perpendicular to V, and inclined to H, and, (c) when it is oblique to both H and V, with two edges JK and ML, parallel to both H and V.

28. **A solid is projected** upon the coordinate planes by projecting points on the surface of the solid; if the solid has plane faces, these points are usually the vertices where three or more faces meet. The projections will be plane figures enclosing, by solid lines, the largest number of points and those edges, which are hidden by the surfaces of the solid, are shown dotted.

FIGURE NO. 14.

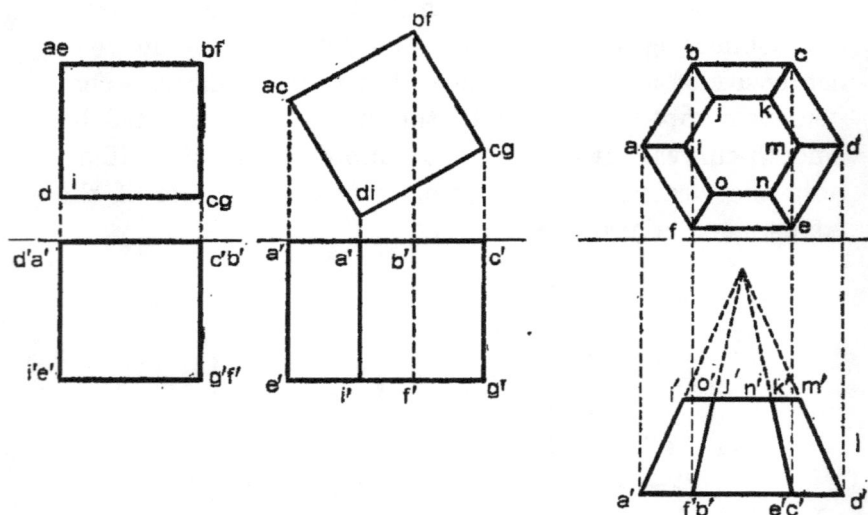

Figure 14 shows a cube in projection, (a) resting on the H plane with four faces perpendicular to H and two parallel to V; consequently, each edge of the cube is projected upon one plane its true length and upon the corresponding plane as a point, shown by the designating letters of its two extremities; (b) a cube resting with one face on H and four faces oblique to V; (c) a truncated hexagonal pyramid. The notation is self explanatory.

29. The end or profile projection of an object is shown in Fig. 15 in the third angle. The upper part gives a perspective of the conditions, the lower, the orthographic projection of the same. Note that the end plane is revolved about its intersection with the V plane and always away from the object. The arrow in the figure helps to show this direction.

30. Any plane may be chosen upon which to project an object; it need not be either vertical or horizontal, but may be oblique to both such planes. It is convenient in case it is desired to show surfaces their full size and shape that are oblique to V and H. The principles of projection are the same and the auxiliary plane, as it is called, is revolved about its line of intersection with V or H into coincidence with one or the other of the latter.

Fig. 16 shows the method of dealing with an auxiliary plane. The upper part gives a perspective of the conditions, the lower, the orthographic projection of the same. The auxiliary plane is taken parallel to the oblique or truncated surface.

31. The circle is projected by projecting points, and connecting these projections by a line, or smooth curve, as the case may be. It is a form of frequent occurrunce in constructive work and deserves to be thoroughly comprehended.

It is convenient to circumscribe the circle with an auxiliary square, putting this into projection and referring the points in the perimeter of the circle to it.

FIGURE No. 15.

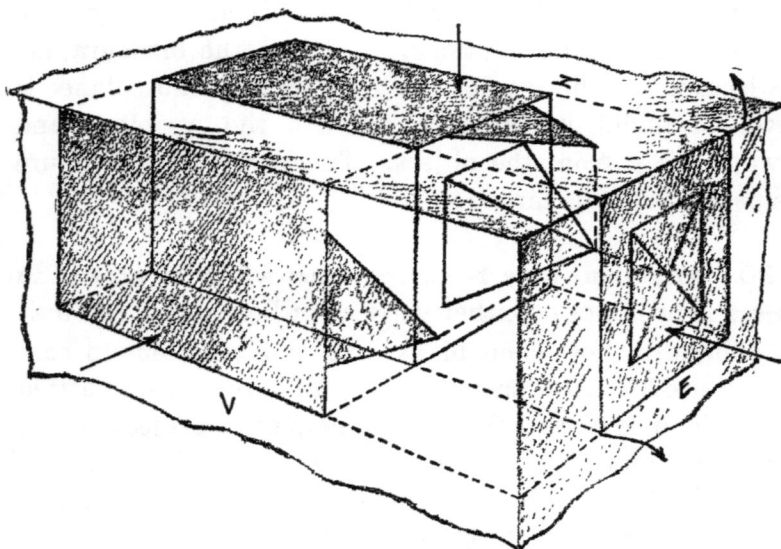

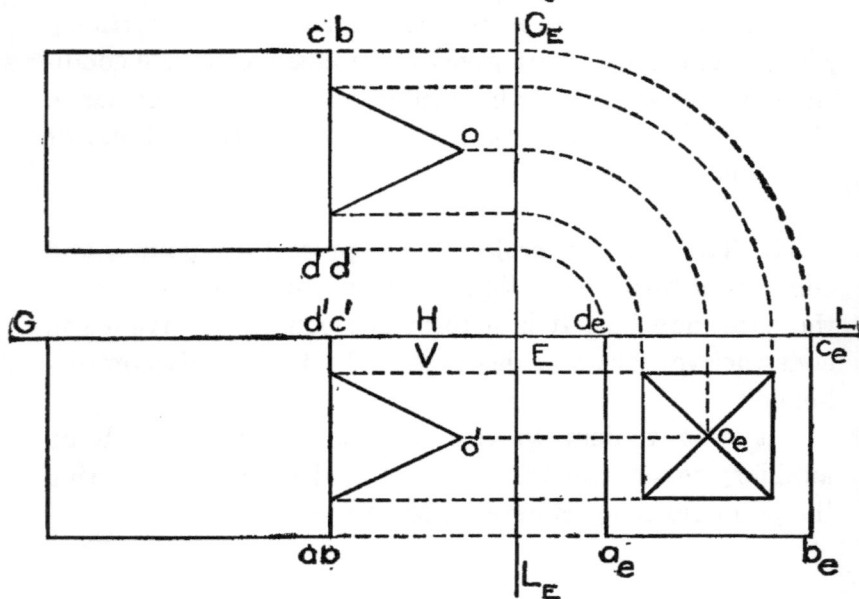

FIGURE NO. 16.

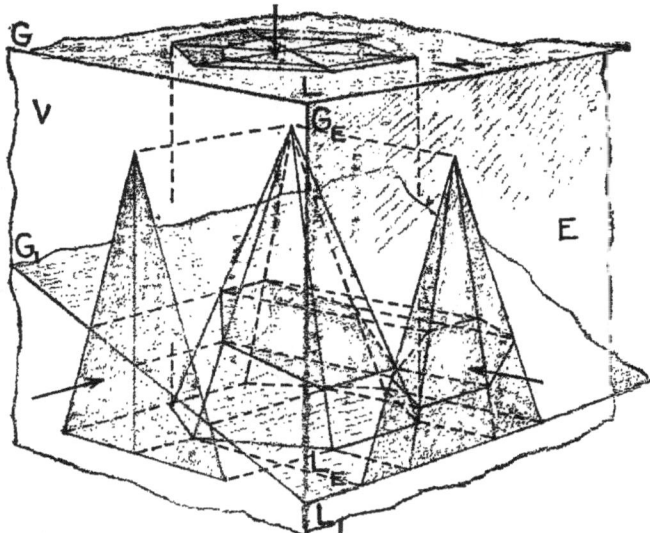

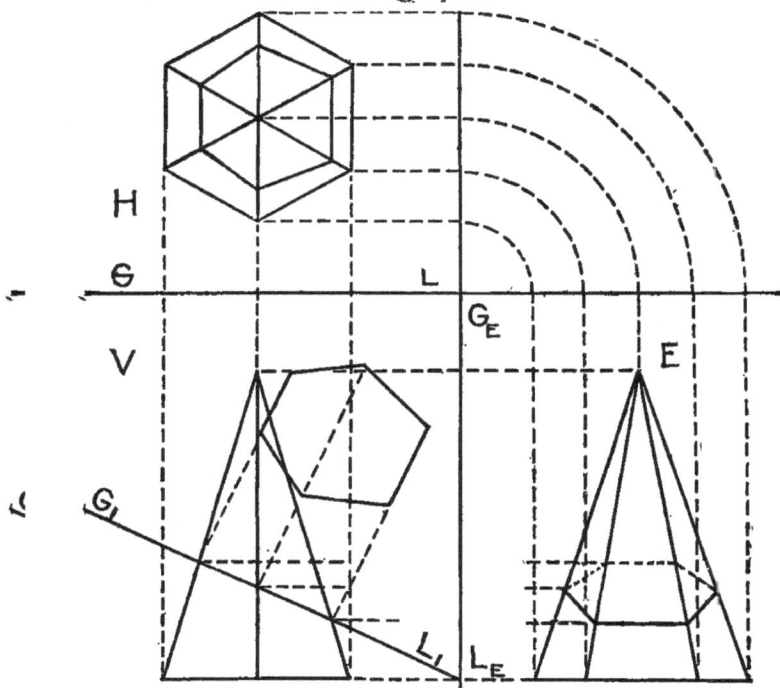

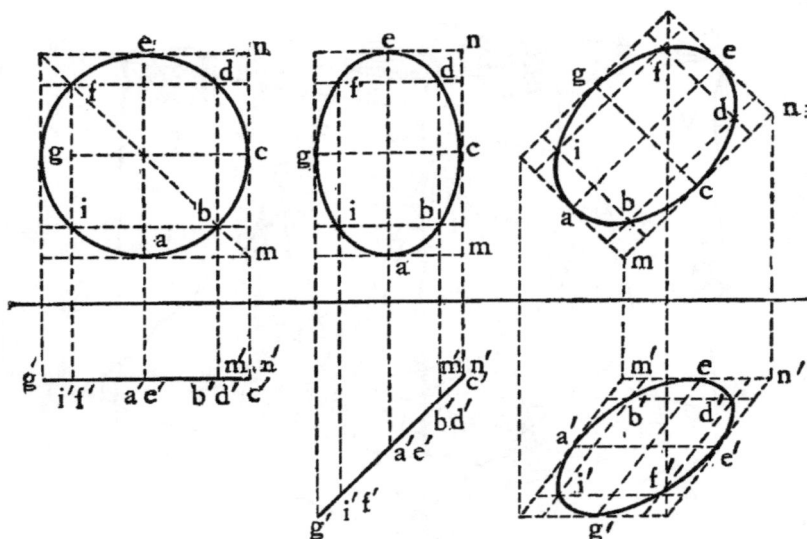

Fig. 17 shows the projection of a circle whose plane is
parallel to H.   In this position its projection on H will be
a circle and, upon V, a straight line equal in length to a
diameter of the circle, as GC.   The points, at which the
circle is tangent to the square, and those in which it cuts
the diagonals, all lie in V projection, upon the line *g' c'*,
as shown in the figure.

Fig. 18 shows the same circle, revolved about the edge
m n of the auxiliary square, until its plane makes an angle
of 60° with H.   Since the plane is perpendicular to V. the
angle of the line *g' c'* with the G.L. shows the angle the
plane makes with H.   The distances of the various points,
in the perimeter of the circle, from the V plane, remain
constant, hence the distances of their H projections from

the G.L. is constant, and they may be transferred from Fig. 17 by parallels to the G.L. intersecting perpendiculars to the G.L., through the V projections of the points, respectively, as shown.

Fig. 19 shows the same circle when its plane is oblique to both H and V. Its projections may be obtained from those of Fig. 18, by conceiving of the edge *m n*, of the auxiliary square, being revolved in a horizonal plane through any desired angle about the point *m*. Since the distances, of points in the perimeter of the circle, from the H plane are constant, the new H projection will not change its shape; and, for the same reason, the new V projection may be obtained by drawing perpendiculars to the new H projections of the points, and intersecting parallels to the G.L., through the V projections of the points, in Fig. 18, respectively.

If a circle is parallel to a coordinate plane, its projection on that plane will be an equal circle, and its projection on the corresponding coordinate plane, to which it is perpendicular, is a straight line. In any other position, the projection of the circle is an ellipse, the proof of which properly belongs to the province of descriptive geometry.

32. Any curve in general may be projected by the principles already discussed, namely, by projecting, severally, points in the curve, or, where convenient, circumscribing the curve by a simple figure, such as a rectangle, and referring its points to the same.

Working drawings, as before remarked, are made according to the principles of orthographic projection that have been discussed. The V projection corresponds to the 'front view,' 'front elevation,' or simply 'elevation,' as the terms are variously used, and the H projection, to the 'plan.' The end elevation corresponds to the 'side view,' 'side elevation,' or 'end view.' The auxiliary projections discussed, correspond to various detail views of parts of a subject, which may have their edges or planes oblique to the principal coordinate planes upon which the subject may be projected.

33. The difference between 1st angle and 3rd angle projection. For working drawings, the choice is open of either the first angle, or the third angle. In either the second or the fourth, the plan is quite likely to fall behind or in front of the elevation. If the distances of the elevation and plan, from the ground line, are made to differ by a sufficient amount, this super-position can be avoided, but an equal difficulty is encountered, in not being able to distinguish which is second angle, and which is fourth angle, for the relation of plan to elevation does not determine it.

In the first angle projection, the object is projected upon the vertical plane, from a center of projection, which is assumed to be on the same side of the plane as the object; this is also true of the horizontal projection. To be consistent, an end view of the object should be that obtained, by projecting it from a center on the same side of the plane as the object. For illustration: The view of the left hand end of an object would be placed on that plane, which was to the right of the object, and the view of the right hand end would be projected upon the plane at the left. Now, if the same center is used, and an object is drawn in vertical projection, in the third angle, it will be projected through the vertical plane, for the center is on the opposite side of the plane from the object. Therefore, to be consistent, the end view should be obtained by use of a center, which is upon the opposite side of the plane from the object. If this is done, the view of the left hand end of an object will lie at the left, and that of the right hand end at the right.

It is convenient to have this latter condition of affairs in a working drawing, because it conduces to legibility, and, in fact, it has been quite universally adopted. Care should be taken by the beginner not to be thoughtless in the use of either angle at pleasure, and in not mixing the two in a drawing.

Figures 20 and 21 show the 1st and 3rd angle projections of an object respectively.

34. *Problems in projection.

(1.) Draw the projections of the following points. (a.) 1¼ inches from H, 1 inch from V. (b.) ¾ inch from V and 1½ inches from H. (c.) In V 1½ inches from H. (d.) In H ½ inch from V. (e.) In the G.L.

(2.) Draw the projections of the following lines, making either projection on V or H not over 1½ inches long. (a.) Oblique to V and H, nearest end to V, ¼ inch from it. (b.) Parallel to H and oblique to V. (c.) Parallel to V and oblique to H. (d.) Perpendicular to H, ¾ inch from V. (e.) Perpendicular to V, and 1 inch from H. (f.) Parallel to both V and H, 1 inch from V

*To be drawn in 3rd angle projection unless otherwise directed.

FIGURE No. 20.

ILLUSTRATING FIRST ANGLE PROJECTION.

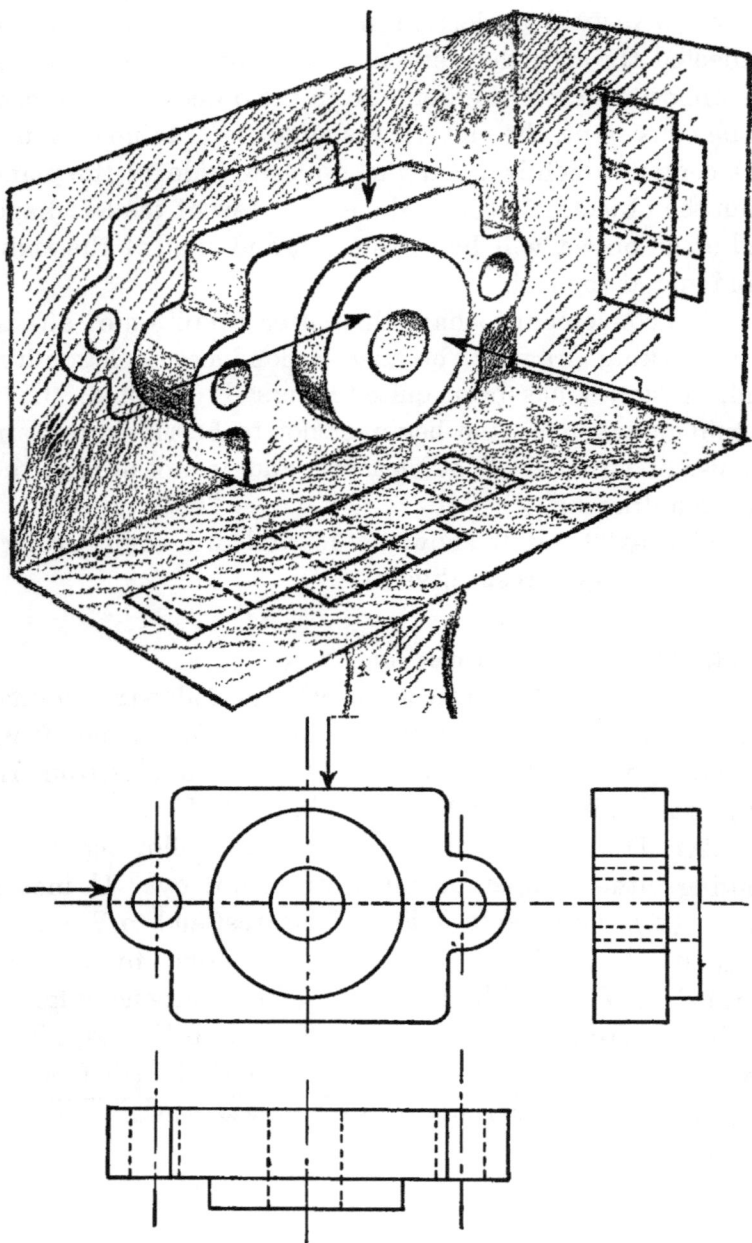

FIGURE No. 21.

ILLUSTRATING THIRD ANGLE PROJECTION.

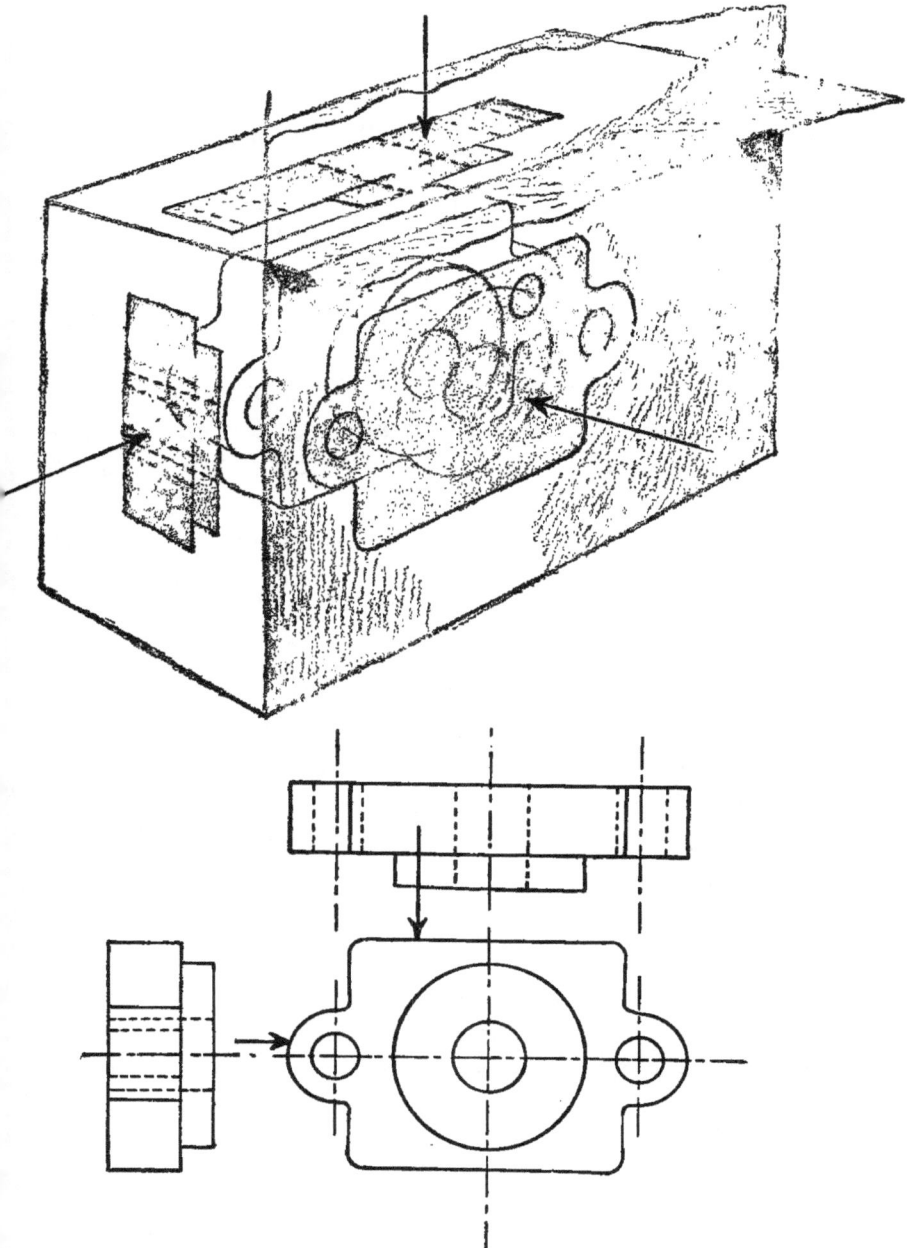

and $\frac{3}{4}$ inch from H. (*g.*) Parallel to and equi-distant from V and H.

(3.) Draw the projections of a rectangle of edges 1 inch x 1½ inches. (*a.*) Place it so that it is parallel to V and distant ½ inch from H. (*b.*) Place it so that it is perpendicular to V and inclined at an angle of 45° to H. (*c.*) Place it so that an edge rests on H, its plane making an angle of 45° with H and the edge upon which it is resting, at an angle of 60° with V. NOTE:—Place it first perpendicular to V, at the required angle with H, and then revolve it into the final position, which will be oblique to both H and V.

(4.) Draw an equilateral triangle in projection of 1¾ inches on edge. So placed that an edge lies in V at an angle of 30° to H, and the plane of the triangle is at 60° to V.

(5.) Draw a circle in V and H and end projection; the circle to be 2 inches in diameter, perpendicular to H, and making an angle of 60° with V.

(6.) Draw the square pyramid, in projection, of dimensions as shown, the base parallel to and 3 inches from H, the edges of base at 30° and 60° respectively to V.

(7.) Draw the pyramid of problem 6 in projection showing also an end view, so placed that the plane of the base is perpendicular to V and makes an angle of 30° with the H plane.

(8.) Draw the hexagonal prism of dimensions shown in V, H, and end projection, with a hexagonal base, resting on the H plane, two rectangular faces parallel to V.

(9.) Draw the hexagonal prism of problem 8, so placed that an edge of hexagonal base rests on H at 30° to

6.

8.

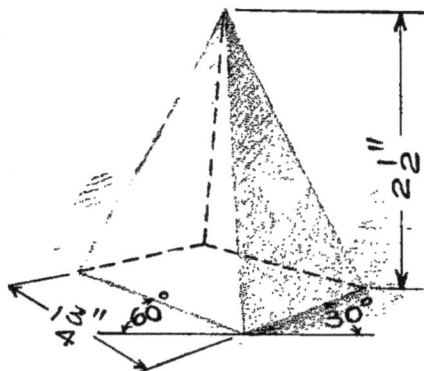

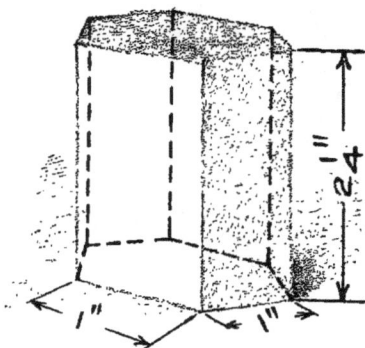

V, with plane of base at 45° to H. Show V and H pro-projections. Project the same upon an auxiliary V plane which is perpendicular to the hexagonal bases.

(10.) Draw the square prism shown, with square base resting on H, and rectangular faces at angles of 30° and 60° with V respectively.

(11.) Draw the square prism of problem 10, placing it so that it rests on a short edge, on H, the edge making an angle of 60° with V, and the plane of the square base at an angle of 60° to H. NOTE:—Draw the prism first with two rectangular faces parallel to V, and then obtain the final position by projecting upon ar auxiliary or V_1 plane at an angle of 60° with V.

(12.) Draw the truncated hexagonal prism shown, so placed that a hexagonal base is in H and all vertical faces oblique to V. The plane of the top is at an angle of 30°

10.

12.

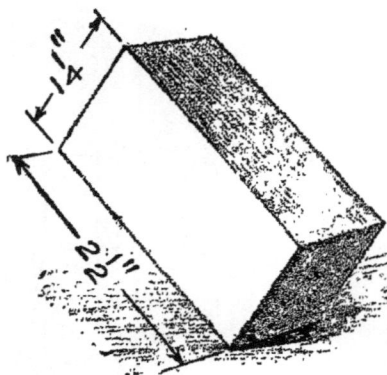

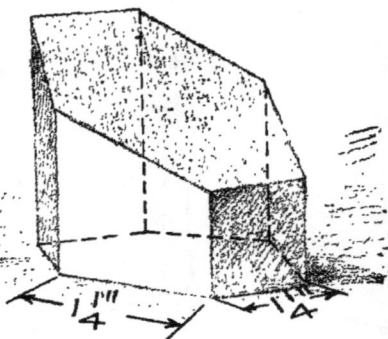

to H. Show the true shape of the section by projection on an auxiliary plane. Assume dimensions not shown.

(13.) Draw the top, front and side views of block shown. Take the dimensions from the figure with the dividers, as they are represented in the picture their true size.

13.

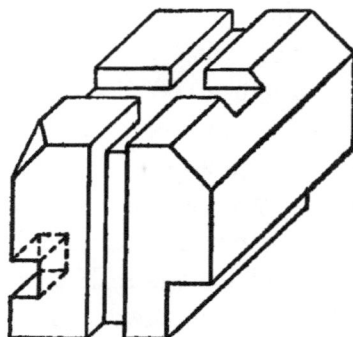

CHAPTER III.

ISOMETRIC AND OBLIQUE DRAWING.

35. Isometric drawing, meaning a drawing of equal measurement, is one which shows form and correct dimensions, within certain limits, in one view. Orthographic projection ordinarily requires two views of an object upon planes at right angles to each other, afterwards revolved into coincidence. A perspective drawing on the other hand, represents, by one view, the object as it would appear from some particular point, giving a pictorial representation, but not the exactness of shape, or the possibility of measurement required in a mechanical drawing, Isometric drawing combines both of these, and shows the three dimensions, length, breadth and thickness in one view and at the same time admitting of measurement. It gives a somewhat distorted appearance to objects, for that reason its application is limited to objects of simple shape.

An isometric d r a w i n g may be derived as follows: If a cube is projected upon a plane, to which all of its faces are equally inclined, the result will look like Fig. 22, and is an **isometric projection** of the cube.[*] All edges, as well as plane faces, are equally foreshortened; and, since

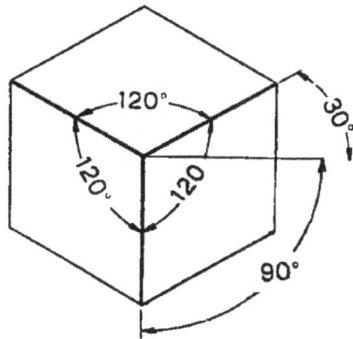

this is true, all edges may be extended until they are equal in length to the edges of the cube, without affecting the proportional relations when we have an **isometric drawing.**

The principles of isometric drawing are derived from this treatment and are as follows:

(*a.*) There are three lines called isometric axes, making angles of 120° with each other. These lines may have any position depending upon the position of the object, but usually those illustrated in Fig. 23.

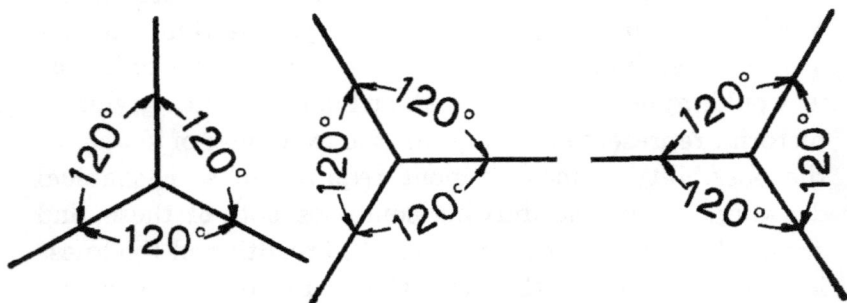

(*b.*) These lines or isometric axes, represent lines which are at right angles to each other.

(*c.*) Upon these lines, or lines parallel to them, the dimensions of length, breath and thickness are measured. All such lines are called isometric lines. Any other lines are '*non-isometric*, and their position and length must be obtained by reference to some isometric lines.

(*d.*) Parallel lines on an object are always parallel on the drawing.

Fig. 24 also shows by the dotted lines how a cube would look in isometric drawing.

Objects, composed of non-isometric lines, must first be surrounded by, or referred to, an object of some

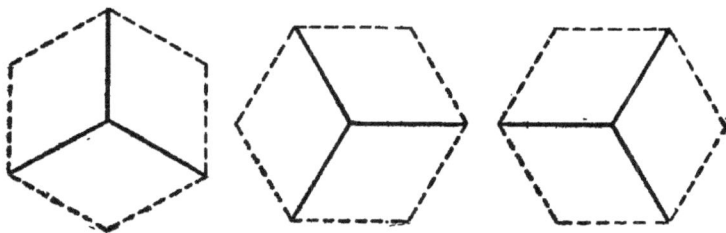

isometric construction, and the lines of the object are then referred to these isometric lines. For example, consider the treatment of a pyramid, see Fig. 25.

FIGURE NO. 25.

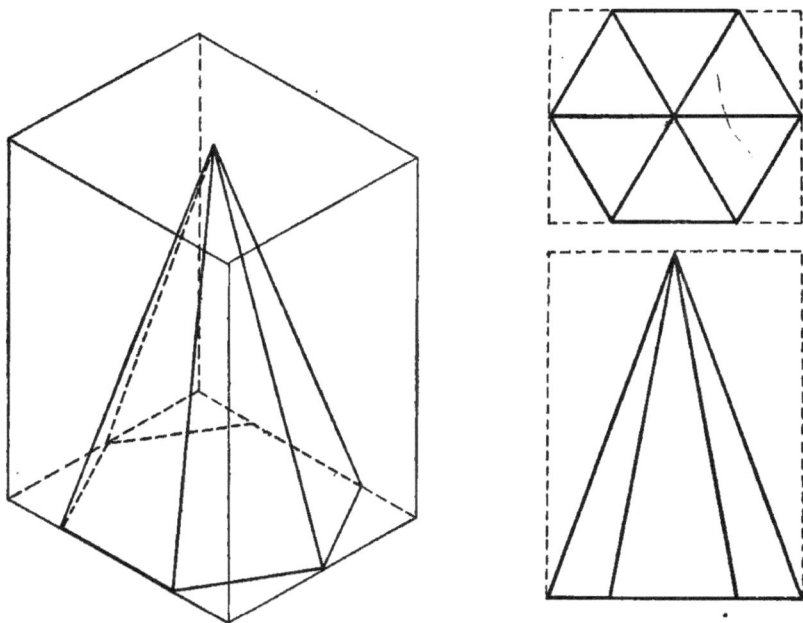

Curves in isometric drawings are obtained by locating a sufficient number of points with reference to some isometric lines, as in Figure 26.

FIGURE NO. 26.   FIGURE NO. 27.

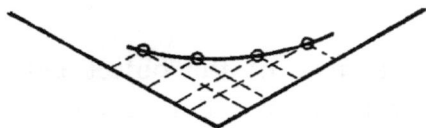
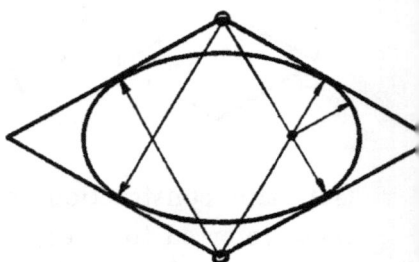

An approximate method for an isometric drawing of a circle is shown in Fig. 27.

36. Oblique projection is somewhat like isometric projection. In most cases its apparent distortion is greater than in isometric drawing, yet it possesses several distinct advantages; the principal one being that two dimensions are represented in their true direction as well

FIGURE NO. 28.

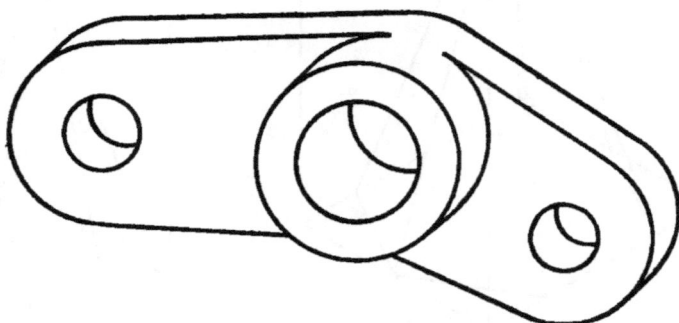

as size, giving one face in its true form as shown in Fig. 28. The principles of oblique projection are as follows: (*a*.) One face of the object is taken parallel to the plane of projection. On this face two dimensions are measured. (*b*.) All lines perpendicular to this face are made 45° lines on the drawing. On these lines the third dimension is measured its true length. (This is not strictly true oblique projection, but differs from it in a similar way that isometric drawing differs from isometric projection.)

CHAPTER IV.

USE OF INSTRUMENTS.

37. Practical points about drawing materials and instruments. A student's first lesson to learn, and a teacher's first duty, is to teach the proper use and care of tools. With this lesson learned, it should be possible for the beginner to work safely with the best appliances, and the best are none too good. It is the purpose now to say something about the different tools and their proper care and handling.

Drawing boards: The best drawing boards are made of well seasoned pine, of uniform grain, narrow strips glued together, the whole being finished at the two opposite ends at right angles to the strips by narrow pieces tongued and grooved and glued to the board, to prevent warping. In large boards of first-class construction, battens are fastened to the back of the board, so as to permit of expansion and contraction of the board with changing temperature, but not of warping. This is effected by fastening a batten rigidly at one point near the middle, and at two or more other points by screws, rigid in the board, but working in slots in the batten. In order to still further lessen the tendency of the board to warp, saw cuts or grooves are made about two inches apart, longitudinal of the strips of which the board is constructed, and of a depth of about half the thickness of the board.

It is not absolutely necessary that all four edges of a board should constitute a true rectangle, they ought to be straight. Only one edge of a board, the left hand edge, should be used upon which to rest the T sq. head, the triangles should be used for vertical lines.

The under side of a battened drawing board may conveniently be used to cut paper on, but it should never be done on the working side, and care should be exercised that the working side and the edges of the board be kept clean and in all respects in good order.

Tee squares are made in various forms. They should be of well seasoned wood of uniform grain. The blade may be sunk in the head or screwed on top and glued. A frequent source of trouble in T sqs. is that the part of the head just under the blade swells under the action of the moisture in the glue when it is made, becomes set, and causes the head to rock against the edge of the board when in use. The bulge in the head can be seen by placing a straight edge against it. An excellent form of T sq. is made of mahogany, with a very narrow edging of ebony. The latter is particularly hard, it relieves the edge by its strong contrast with the color of the paper. Sometimes a celluloid edge is used. This seems to have growing favor because of its transparency, permitting partial sight of the work underneath. If T sqs. are large, they are tapered as before mentioned, so that the upper edge is the only one that can be used.

Some T sqs. are made with a swivel head so that the angle of the blade can be adjusted, and they have their value. It may be said that one should be in a large and well equipped drafting office, but that for ordinary use it is not necessary. Steel T sqs. are made, but are not favorites with draftsmen because of their weight and the danger of injury to the drawing by denting, etc.

Triangles are made of the same materials as the T sqs., solid triangles, however, are not good as they will warp. Triangles are made also of vulcanized rubber and of celluloid. The former have the advantage of contrasting well with the color of the paper, but they have the disadvantage of being non-absorbent, consequently they transfer dirt from one part of the drawing to another and are apt to smear it. The inside edges of a triangle may have depressions cut in them to facilitate picking up from the paper.

A 30° and a 60° triangle and a 45° triangle are the only two in common use except those especially made for mechanical lettering. Some other angles may be struck by using the triangles together and adding or subtracting their angles to get 15°, 75°, etc.

Paper comes of various kinds and quality, suited to different kinds of drawing. A moderately heavy grade with smooth, hard surface is to be desired for mechanical drawings. A yellow or manilla paper is much used for pencil drawings, when the resulting drawing is to be traced; it is called detail paper. Bond paper has also come quite into use, and it has distinct advantages. The drawing is penciled and inked on the paper and from it blue prints can be readily made. Paper should, if possible, never be rolled, particularly rolled small as it cannot be satisfactorily flattened again.

Tracing cloth is in almost universal use in drafting rooms for permanent drawings, as blue prints can be so readily made from it. Either the rough or the smooth side can be equally well used for drawing in ink. Frequently the original pencil drawing is made upon the rough side of the cloth and inked over. It furnishes a very good surface for this purpose. The smooth side is impractical for pencil drawings but takes ink like a highly calendered surface. Special precautions, which will be mentioned shortly, have to be observed in working upon it. There is but one recognized grade and make of tracing cloth, the "Imperial."

Stretching of paper, as it is called, is resorted to where any water color brush work is to be done. It consists of pasting the edges to the board and shrinking until it is quite taut. It takes a little experimenting to get facility in doing this, but every one ought to know how to do it when occasion arises, hence the following directions are appended.

* "To stretch paper tightly on the board, lay the sheet right side up—which side is presumably the one which shows correct reading of the water mark when held to the light—place a rule with its edge about one-half inch back from each edge of the paper in turn, and fold up against it a margin of that width. Then thoroughly dampen the back of the paper with a full sponge, except on the folded margins. Turning the paper again face up, gum the margins with strong mucilage or glue, and quickly but firmly press opposite edges down simultaneously, long sides first, exerting at the same time a slight outward pressure with the hands to bring the paper down somewhat closer to the board. Until the gum sets so that the paper adheres perfectly where it should, the latter

---

* F. N. Willson's Theoretical and Practical Graphics, p. 14.

should not shrink; hence, the necessity for so completely soaking it at first. The sponge may be applied to the face of the paper provided it is not rubbed over the surface so as to damage it. The stretch should be horizontal when drying, and no excess of water should be left standing on the surface; otherwise a water mark will form at the edge of each pool."

The compass, dividers, bow instruments and ruling pen constitute the simple, universal kit, and probably the majority of draftsmen have little else. Of course there are a number of other tools made, chiefly for special uses. These are not used universally, however, because the time saved with a special tool is usually offset by the time consumed in handling and cleaning, for each special tool comes in generally for but occasional and brief use. Somewhat similar reasons explain why various special attachments to the simple kit are not popular universally, like hair spring legs in compass and dividers, spring catch ruling pens, micrometer adjustment to needle point, etc.

Beam compasses are instruments to strike large circles, consisting of a needle point leg and marking point leg, each separate and adjustably mounted upon a bar of metal or wood. Every large drafting room is likely to have one for occasional use, but the individual hardly needs to go to the expense of one unless its use is demanded frequently.

Follower pens, in which the pen is swiveled in the handle, are used to make irregular curves. The pen automatically adjusts itself properly to the ruling edge. It has but occasional use.

A bow pen, made chiefly for special professions, has a fixed needle point leg with a marking leg sliding freely upon it. It is handy for striking a large number of circles of small diameter, but it is a tool for that special purpose.

Dotting Wheels are instruments to do what the name implies, make dotted lines. They also have occasional use but are a trouble to care for and easily get out of order.

Proportional dividers, consisting of double pointed legs, pivoted between the ends, and adjustable, so as to give a range of relation between the opposite angles formed, are a very useful tool indeed upon those rare occasions when a drawing is merely to be copied to a different size regardless of scale. Where scale is desired it is not safe as a tool nor is it much handier than the scale direct.

A parallel straight edge is made which replaces the T sq. A rule, of the length of the board, is held at the ends by sliding on a wire cable and moves into parallel positions. Theoretically it is excellent but lack of sufficient rigidity is its chief drawback in the opinion of many.

The protractor is a semi-circular disc segment of celluloid, bone or brass with degrees marked upon it. The center is marked on the straight edge of it. It has use where angles have to be struck of varying sizes and other than those for which the triangles can be used.

There is a machine on the market known as the **Universal Drafting Machine,** which has very meritorious features. It combines the function of the T sq., triangles and scales, and, when specially adjusted, the protractor.

It consists essentially of two straight edges with scales upon them and with a common point of attachment. They can be set rigidly at any angle to one another and the whole moved in any direction over the board through the medium of hinged arms, rigidly attached to the upper left hand corner of the drawing board. Straight edges, having any of the standard scales upon them, may be attached to the frame.

There is also a **Paragon Drafting Instrument** accomplishing much the same purpose. It is attachable to a parallel ruler previously mentioned or it can be attached to a T sq. blade. The fixed center is the point of attachment and the ruling edges, two in number, can be swung around it at any angle, replacing triangles and protractor, and also having variously scaled edges.

38. Some practical points about and the care and handling of drawing instruments.

Drawings can be cleaned of dirt with the soft pliable

erasers, the kneaded rubber, the sponge rubber or stale bread crumbs rubbed over with a cloth or with the hand. The liquid drawing inks will stand very little erasure with the pencil eraser without loss of blackness in the lines. To keep a drawing in good shape as the work progresses, cultivate early the habit of keeping the T sq. and triangles clean, using a piece of paper where possible over parts of the drawing not in immediate use, and, finally, *keeping the hands off the work* when they are not in active service.

Be careful to use the tools only for the purposes for which they were intended. Violations of this are to be found in using the T sq. as a hammer to put in tacks, the dividers as compasses to describe arcs, sticking the divider points into the board so the dividers will stand alone, etc., all of which tend to injure the tools.

The tools should be at all times handy. With the T sq. always on the board, the triangles above it on the board and other tools in predetermined places from which they can be picked up without much, if any, hunting, and while the eyes are engaged on the drawing, will conduce to rapidity and accuracy of work. Observe that the workman in any craft will always lay a tool down when he is done with it, even temporarily, and moreover, he lays it down where it is the least trouble to find it again.

Facility in the use of the ruling pen is eminently desirable, hence a few more practical directions are here given:—

The greater care at all times should be exercised, the thicker the line used or the fuller the pen is with ink. The beginner should carry less ink in the pen than after he becomes an expert. When occasion arises to use the pen for long lines or many close together, with the least interruption for refilling, a considerable amount of ink can be carried in the pen if the following directions are observed: Head the pen to start, the point very close to but not touching the paper; when ready, touch the pen to the starting point and instantly move on the line uniformly and rapidly, the more rapidly the fuller the pen. Stop and lift the pen in the same instantaneous way. These same precautions hold when making a very thick line with the pen; the thicker the line the less ink can be carried.

In drawing lines to go from or to a heavy ink line or border still greater care has to be observed that the border does not draw the ink out of the pen and cause a blot.   The

FIGURE No. 29.

situation is illustrated in Fig. 29. It shows a series of lines close together where the edges of the lines are apt to break down and the lines run together. In cases of great danger, every other line or so may be begun late as shown in the figure and afterwards filled out when the ink dries. In patching these open spaces set the pen to make a finer line, matching only one edge of the line drawn, then by tilting the pen probably the requisite increase can be made if not with accuracy in one stroke, then in two or more; a line can be added to easily but it cannot be reduced in size, except by erasure first and then redrawing.

If a pen fails to work it may be due to several causes

(*a.*) It may be set so tight that the ink cannot flow out between the nibs.

(*b.*) The ink may have dried at the ends of the nibs, if not farther, and clogged the flow. The best thing to do is to at once clean and refill. The use of the blotter or a piece of paper drawn through between the nibs is to be deprecated.

(*c.*) If the ink does not run down to the point, proper running may be facilitated by opening the nibs a little and shaking gently, over a blotter or something it will not injure if it blots, until the ink settles or drops out. The difficulty is caused by grease on the inside of the nibs.

(*d.*) The pen point may be actually out of order. This of course demands that it be sharpened; but the pen should be tested for all the other difficulties first. An injured pen will either not mark at all or it will make a ragged line; the line, moreover, may be ragged at one or both edges. If the pen is merely uniformly dull, it will refuse to make a fine line, the line will simply fail entirely if the nibs are brought close together.

A pen can be sharpened and tested in the following manner: A fine oil stone should be used for this, an Arkansas stone seems to be preferred. Bring the nibs of the pen together as for drawing a very fine line, and hold for the rubbing at a small angle to the stone, 30° or less, and with the broad face of the nibs towards the stone. Rub to and fro in the direction of the handle with at the same time a slight rocking of the pen in order to round the point. If too pointed it tends to cut into the paper and will not hold sharpness so long. To test for sharpness, drag it on a piece of paper as if making a line; it ought not to scratch roughly or glide too freely, but bite slightly, that is, resist motion. If it seems to act as it should, clean thoroughly and then try with ink. Properly sharpened, the pen should make a very fine black hair-line without breaking and a broad line of sharp edges, even if the pen is tilted five or six degrees out of plumb in a plane perpendicular to the ruling edge. Try for a broad line first with this test in order to see if both sides are of equal length. If they are not, that side of the line at which the nibs are shortest will show ragged. If this test is successful and the line drawn is perfectly sharp and clear on its edges, test for fineness of line, by working from a wide line towards a narrow one. If it happens that the sharpening has proceeded too far and the pen bites too deeply into the paper, or if one nib is slightly longer than the other, the pen may be dulled or the long nib worn down by rubbing it on the stone with a rotary motion when the broad nibs of the pen lie in a plane perpendicular to the plane of the stone.

To determine the place of a line the ruling edge should furnish a rough approximation and the marking point the exact place. It is one of the important points in handling to be learned early. The nibs of a ruling pen, for example, being bowed, will touch the paper slightly beyond the ruling edge. If the pen is incorrectly tilted until the nibs touch the paper at the ruling edge, a blot is almost sure to result, for the ink will touch the ruling edge.

It may sometimes happen, when a number of lines have to be drawn which run in a variety of directions, that waste of time is threatened in waiting for ink to dry. The ruling edge can be held slightly free of the paper and over the wet lines by using the thumb and first or second finger as a cushion underneath it or one ruling edge may be rested on and slightly overhanging another. In small work one triangle may be put with its open space over the lines to be drawn, and the other triangle rested upon it, crossing the gap. A method of inking will be shortly discussed which overcomes some difficulties of waiting for ink to dry.

Errors in an ink drawing can be corrected so that the repairs are practically invisible. A knife will not do this unless used in conjunction with the ink eraser. Iu fact, cuttiug or scratching with a knife is so risky that it is safe to adopt the custom of never using it except under extreme circumstances.

If an error occurs, take up as much ink as possible with a blotter, but do not use it under any ordinary circumstances to dry a line because it pales the ink. Then use the ink eraser, rubbing rather lightly and rapidly, not in one direction or with one part of the eraser, but in all directions and changing the point of contact, because the rubber will heat and not work so well. Every vestige of the mistake should thus be removed, although it blurs a certain area around the error. Next clean off all the sand by using the pencil eraser. If the surface of the paper is very much disturbed it may be necessary to burnish it with a piece of ivory or smooth metal. The difficult part of correcting errors comes in putting back the ink lines. A line made upon an erased space is quite apt to spread and show larger than on the fresh paper, although the difference is very slight, therefore the pen should be set for a slightly finer line, and this added to by successive strokes. If, in spite of all precautions, the place erased be treacherous, use two exceedingly fine lines as limits or walls, the distance apart of the thickness of the line to be drawn, and which, when dry, will prevent the filling ink from percolating into the rough paper. In case of a very wide line, the retaining walls may have to be built up gradually. In repairing also it is necessary to overlap the correct part of the line sufficiently to include all that has been affected by the erasing. A very good hard drawing paper ought to permit several, say three or four, corrections over the same spot if skillfully managed. Corrections upon the rough side of tracing cloth are very easily made with the ink eraser and no burnishing is necessary. On the smooth side, however, erasing is difficult and quite apt to irreparably injure the surface of the cloth. The greatest of care must be used by rubbing lightly to prevent trouble from this cause. A knife is almost sure to take off the surface, and if it does, burnishing will not repair the injury.

A knife comes of service, now and then in one of two ways, first to scratch off the crust of large blots or very wide lines, without attempting to remove the ink entirely; second, to cut out an extremely small spot of ink or a slightly overlapping line. In the case of the latter, the knife should be run along the edge of the correct portion to cut it away sharply from the incorrect, then the error may be scratched free without leaving the correct line ragged.

39. General directions for penciling drawings.
The size of plates 12 inches by 18 inches, outside.
Border line $\frac{1}{2}$ inch from outside edge.

Tack the paper by the upper left hand corner, then, with T sq. head against the left hand edge of board, swing paper into line with its upper edge. Next, drawing tight, tack upper right hand corner, then the lower two corners.

If the drawing has to be temporarily removed, draw short horizontal lines on each side of the sheet and extending onto the board, to guide in replacing.

Hold the T sq. with the hand over the head or by the blade close to the head. With the latter way the blade can be made to creep by means of the fingers for short distances across the paper.

Keep the T sq. against the left hand edge of the board, and use only the upper edge for working against.

Keep the triangles convenient to the T sq. and, when through using, move to the right or upward from the blade.

In using a triangle against the T sq. observe the following method: Adjust the T sq. first with the right hand, bring the triangle into place and hold with the fingers of the left hand while the ball of the hand rests on the blade.

FIGURE NO. 30.

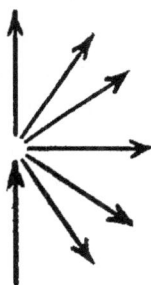

Rule all lines in pencil and ink in the directions shown in Fig. 30. The vertical lines should be ruled against the left hand edge of the triangle. Do not draw to the extreme point of the triangle. Lines locating points should cut each other as nearly as possible at right angles.

The lead pencil may be sharpened in one of two ways, by a long, tapering, round point or by a double-edged chisel. Cut the wood back for at least $\frac{3}{4}$ of an inch from the end and leave from $\frac{1}{4}$ to $\frac{3}{8}$ of an inch of lead exposed. Taper both down continuously to, if possible, a slightly concave form. The advantage of a tapering point is that it holds its sharpness for a longer time, and again, the point is not thick enough to cover up the work in hand or to mislead as to where the lead is marking.

The double-edged chisel should not be quite as wide across as the lead is thick, but reduced somewhat, say to $\frac{3}{4}$ the diameter. A penknife for the wood and emery paper or a file for the lead will give the desired results most rapidly.

If the double-edged chisel is used it should only be for straight away lines, not for laying off measurements from the rule or scale.

Hold the pencil nearly perpendicularly to the paper; if drawing lines with the round point, acquire the habit of slowly twirling the pencil during motion, so that the point will be worn down in a conical shape, and not irregularly.

Clean, firm lines, uniform in thickness and blackness both in penciling and inking, are the kind which should be cultivated.

Avoid drawing superfluous lines, lines overrunning their proper limits or lines that are not to be inked. If accidental errors are made, correct them with the eraser at once.

Lines, which are to be dotted in final drawing, should be dotted in pencil so that no mistake is made when inking.

After making an erasure, clean off the particles of dirt that are loose on the paper, for they interfere with the smooth and proper action of the other tools. This is a particularly important direction to observe preparatory to inking and when any alterations are made during inking. To much care cannot be observed in freeing the paper and all tools from the dirt particles, for they are quite apt to get into the pen and give trouble.

Successful work of any kind must proceed in a systematic and orderly manner. A system in penciling cannot be followed to advantage entirely, because the conditions in the development of a drawing vary quite a little, nevertheless, a general plan can be followed when circumstances permit. The following is such a system:—*

System in Penciling.

1. Draw border lines.
2. Draw match lines to guide in replacing the drawing, if it is temporarily removed.
3. Block out space for title.
4. Block out space for bill of materials.
5. Block out the views to be placed upon the sheet.
6. Draw main center lines, and where these are to be inked, they may be drawn full light lines.
7. Locate main lines of views.
8. Draw small and inside lines.
9. Put on dimensions and necessary notes.

40. General directions for inking.

To put ink into a right line pen, hold the pen approximately horizontal with the kind of holding usually given to a writing pen. Hold the bottle down with two fingers and with two fingers lift the cork out and touch the quill end between the nibs of the pen, to let the ink run in, and do this over the bottle, corking it securely when through. If the nibs of the pen get ink on the outside, wipe off with a rag.

Hold the pen perpendicularly to the paper, steadying the hand against the ruling edge with the last finger or the last two fingers. Place the first finger just above the

---

*Coolidge & Freeman, "Mechanical Drawing."

adjusting screw on the flat part of the nib, the thumb just opposite, and the second finger touching the pen between the first finger and the thumb and just below them to steady it from any tendency to turn.

Bear no weight on the pen, its own weight should be sufficient to make the desired line. If it does not, clean out and refill.

Do not press pen against the ruling edge, as it will tend to close the nibs.

Make all lines by a continuous motion of the pen.

Do not stop on a line to see where the rest of the line is to go. If it is absolutely necessary to stop in any case, even for a moment, lift the pen from the paper. Also, when stopping in this way, take the further precaution to move the ruling edge away from the wet line.

When requiring refilling, clean the pen out thoroughly inside and out. It is well to cultivate the habit of doing this early, making it invariable, for a pen clogged with ink is likely to give trouble. Use the nails of the thumb and first finger successively, covered by one thickness of rag, and·it will be found easy to clean one-half of a blade with one finger and the other half with the other, and the remaining blade by turning the pen over and repeating the operation. Four wipes generally clean sufficiently.

If the pen gives any trouble in marking at any time it is safest to empty, clean and refill.

The weight of an ink line on a drawing should be such that it will show clearly the form within the maze of dimension lines, etc., that it will blue print readily, giving an equally clear impression throughout, and that all annoyance is removed, due to the likelihood that the line will

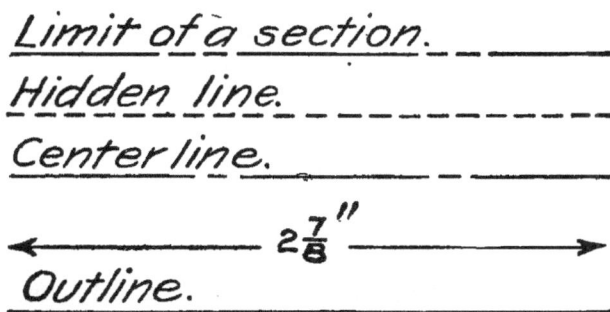

Limit of a section. _ _ _____

Hidden line. ___ _ _ _ _ _ _ _ _ _

Center line. _____ _ _ ____

← ———————— $2\frac{7}{8}''$ ——————— →

Outline. _____

break, through any obstructed flow from the pen. In Fig. 31 are shown suitable conventions to use on a drawing, as well as the proper weight of line.

System in inking is more imperative than in penciling owing to the trouble of changing tools and waiting for ink to dry. A general plan which will be found useful follows:

System in inking.

1. Ink all small circles and arcs of circles with the bow pen.
2. Ink larger circles and arcs with the compass.
3. Ink irregular curves with curve ruler.
4. Ink all horizontal lines with the T sq.
5. Ink all vertical lines with the triangle resting on the T sq. edge.
6. Ink all 45°, 30° and 60° lines in groups and in order.
7. Ink other oblique lines not at the above angles.
8. Section lining.
9. Dimensioning.
10. Surface tinting and shading.
11. Lettering and descriptive matter.

In large complicated drawings, treat a small portion of the sheet at a time complete with one tool until the whole has been covered. Some authorities advocate drawing center lines first, and it is also a good method.

Section lining and dimensioning may conveniently change places in the series where dimensions do not have to be written across sectioned surfaces.

41. **A drawing should be cleaned** after all inking is finished. A soft rubber will clean off the dirt well, but it is not sufficient for erasing superfluous construction lines. Take the latter out carefully with a harder eraser so as not to injure the ink lines.

42. **Handling of the compass, dividers and bows.**

The compass is for describing circles and measuring angles, and also for transferring measurements from one place to another.

The dividers are used to approximately subdivide linear distances and for transferring measurements from one place to another. It is, many times, a more convenient tool for doing these things, and one of the habits to cultivate is to minimize the use of the dividers. It is an excellent tool in its place, but it is not as safe to depend upon as the scale.

In changing the marking legs of the compass, use care to pull or push the attachment longitudinally of the leg, and not to twist it or move it laterally, it might strain the members. The instrument is easily injured, and accuracy of action is necessary.

The needle point of the compass should have a shoulder

on it to prevent sticking too deep in the paper; it should be adjusted to fit the pen attachment and always be kept so. In working with the pencil leg then, and as it wears down, extend the lead to meet properly the needle point adjustment. The pencil in the compass should be sharpened always to the double-edged chisel.

The proper position for the needle point is slightly in advance of the marking point, depending upon the degree of sharpness and the length of the point. At no time should the point more than hold in the paper. And when stuck in to this degree, the bisector of the angle of the compass legs should be perpendicular to the line connecting the two points. This is very important in making small circles.

The compass should be held with one hand only, for reasons apparent after experience.

To open the compass at first, press the thumb and first finger against the bevelled portion near the head.

To hold the compass when opened, control the needle point leg with the thumb and third finger, the marking leg with the first and second fingers, held generally, the former upon the outside, the latter upon the inside of the leg, so as to move the leg in and out with a controlled motion.

The tightness of the head should be just sufficient to hold the compass in place during use; such an adjustment will not render it difficult to change the angle between the legs easily with the fingers as described.

If the hand is unsteady, it may be found convenient to use one hand for putting the needle point in the proper center, either by taking hold of the end of the needle point leg or by resting the leg against the finger while putting it into place; the latter is the better way.

Curves should be drawn continuously and always clock-wise. In drawing a complete circle, start the marking point at the lowest part of the circle, or even a little to the right, and it will be found possible to swing the whole circle without change of handling, by rolling the head in the fingers during rotation. It should not be necessary to change the handling at any time, or the position of the hand on the instrument. Held correctly, it should always be ready for drawing. It is a very common fault to hold the compass with two hands.

It should be held so that the head is slightly in advance of the marking point in the direction in which the curve is being made, to aid the marking point to remain in contact with the paper. But do not bear any more pressure on either leg than is necessary to make the curve and to keep the needle point in contact with the paper; in ink work, the weight of the instrument, as in the case of the ruling pen, should be sufficient to make the desired line.

Do not overlap a circle in inking; there is a chance of a change of adjustment, and even if this is not the case, it is likely that a line twice drawn over will spread out making a noticeable junction.

The hairspring attachment put on some compasses and dividers has merits in enabling one to make a very delicate change of adjustment, but the author thinks the value of this feature is very much over estimated, for, with experience, comes sufficient skill that, handled as above described, the desired adjustments are made more rapidly than they could be by a hairspring attachment. Of the two instruments, however, the hairspring is of more value upon the compass than upon the dividers.

For large circles, bend the legs of the compass at the joint provided so that they come down perpendicularly to the paper.

A lengthening bar is used for circles beyond the capacity, ordinarily, of the compass, but it is an inconvenient thing, makes an unsteady tool, and if much work is to be done on large radii it is well to use a beam compass, built especially for this purpose.

The dividers are held in the same manner as the compass. If a given distance is to be divided into a certain number of equal parts that do not correspond to any scale divisions, the dividers can be used to do it by successive approximations. In stepping off such equal spaces, the tool should be swung alternately over and under. Moreover, but slight pressure should be exerted on the tool, so that the points make no noticeable hole in the paper.

To locate a pick of the divider leg for further reference, put a small free-hand circle about the point, not much over a sixteenth of an inch in diameter; this calls attention to the region in which the point lies. Large holes in a drawing are unsightly and are really inaccurate.

The bow instruments are very convenient, small and accurate tools for doing the same kind of things that the compass and dividers will do. The adjustment of needle to marking point in the bow pen and pencil should be even more carefully made than in the compass, because of the small circles for which they are used. On account of their accuracy and positive adjustment, the bows in practical work are used wherever they can be, but since there is nothing distinctive to be learned about them, the beginner is advised rather to favor the use of the latter so that he may get as much practice with them as possible, and acquire the proper handling, which does not generally come naturally at first.

The small circles upon commercial drawings are not infrequently omitted in penciling, only the centers being located, but a caution is extended to the beginner not to resort to this form of short cut, but to pencil in everything very completely and accurately. The omissions of construction should be left to the judgment of the skillful draftsman.

If a decided change is to be made in the adjustment of the bows, it is less wear on the instrument and also more economical of time to take the strain off the legs by pressing them together with one hand while the thumb screw is twirled around with the other until near the proper adjustment.

43. The use of the irregular curve: The irregular curves are those which cannot be drawn accurately with the compass. They must be plotted by points and the latter joined by a smooth curve made with the curved ruler or "curve" as it is called.

The curve can be drawn best at first free-hand and then copied with the "curve."

The correct shape for a "curve" is of importance. The best is one which has the fewest and simplest curves in one tool, as the spiral form for example.

The "curve" should be applied to the points plotted with the direction of the change of curvature the same in both.

The drawn curve should be as unbroken as the theoret-

ical curve, and its execution is not easy at first. To insure matching the curve, apply the "curve" to at least three points, more if possible, then draw, not as far as the ruler seems to match the curve but a little short of it, the amount depending upon how rapidly the ruler departs from the curve to be drawn at the last discernable point. Again, in moving to the next segment, match the ruler to the part already drawn so that it corresponds with it for an appreciable distance back of the last point drawn to.

To connect the segments of the line accurately, head the pen first to make a perfect alignment, then start to move on the line, or if drawing up to a line, and just before reaching it, tilt the pen, if necessary, to bring it into the ink line correctly, but do not overlap.

Keep the blades of the pen at all times tangent to the ruling edge, but do not work on the under side of the "curve."

CHAPTER V.

WORKING DRAWINGS.

44. **Orthographic projection and working drawing.**

Orthographic projection is the language in which working drawings are written, but a dimensioned orthograghic projection of anything does not necessarily constitute a working drawing of that thing.

A working drawing must be simple and plain in its features, easy to be interpreted, yet explicit. For if one view of a piece will suffice to tell a workman how to make it, only the one view need be made. On the other hand, however, more views may be required in the working drawing than are required in orthographic projection, for sometimes assembly views are needed to show relation of all parts and detail drawings of each component part in addition. The principles of orthographic projection are frequently violated in working drawings wherever modification will aid in legibility or economy of time in drawing.

There is no way to formulate this difference between the two under rules for there are no fixed ones. The draftsman should place himself in the position, in imagination, of the one who is going to construct from his drawings, and in that way arrive at a conclusion as to what would be desirable in the way of views. Unless the draftsman does this, he is apt to economize time and effort at the expense of the workman's time. From the workman's standpoint, many times, drawings are not explanatory enough; he will want them too elaborated with directions. A mean of these two has to be struck.

45. A set of working drawings in its completest form, consists of diagrams, assembly views, details, sections and bill of materials.

A diagram is a drawing which is first made to determine upon the arrangement or lay out of the various parts. Upon this lay out, also, may depend the character of the forms, so that this is an additional requirement for its being made first. The diagram shows, further, the number of the various elements of the group. As one illustration of this kind of drawing can be mentioned a layout for piping, showing the number of elbows, tees, valves, etc.

In a diagram the briefest indication of shape is given, and not infrequently special conventional forms are used to stand for the more intricate actual forms. In piping, for instance, a valve is represented by two short lines perpendicular to and crossed by each other, one perpendicular to the line of piping and standing for the entire valve, the other parallel to the line of the piping and standing for the handle.

Another illustration of the diagram is an outline of a machine composed of the main lines, together with the usual center lines. Perhaps, in this, the relation of some of the moving parts is shown. These may be represented by heavy lines coinciding with the center lines of the members for which they stand, as in a Corliss engine valve gear diagram.

The assembly drawing shows the entire subject to be

treated. It may not show all the features, or parts, only the principal ones; but it gives certain facts not available in any other way. It shows the size of the whole, the place for the different component parts, and the relation between these, together with certain desirable chief dimensions. The minor features, such as bolts, nuts, keys, set screws, etc., are left off. Perhaps their place will be indicated by center lines; perhaps, not even that. In fact, the assembly drawing may be more or less in the form of a diagram.

The details are made together upon a sheet of details, or each may be made on a separate sheet for the different workmen, according to the process through which the parts are to be put. There are details, for example, for the pattern maker, the blacksmith or the machinist. The dimensions put on these and the general treatment will be that of interest to the particular workman handling them. Sometimes the detail drawings are made complete enough in all respects to answer for the several above mentioned requirements.

46. **Sections** are made many times to save the drawing of details. A section in its simplest form, means to cut anything as with a saw and to show by some conventional, or commonly understood means, the plane of the cut and what lies beyond this plane when looking perpendicularly at it. The treatment of sections will be taken up a little later.

47. The bill of materials is a tabulation of the stock required, the number and character of the pieces needed.

To be specific, it is composed of: (*a.*) An identification mark as a number, which may, by its denomination, indicate the material. (*b.*) Name of the part. (*c.*) Number of the pieces needed to make one of the entire subject. (*d.*) Name of material, if the identification is not complete as above. (*e.*) Further general descriptive matter, like pattern number, dimensions of the rough stock, method of casting, etc. In very small subjects it may not be made a separate tabulation, but written near the separate parts. In other cases it may be a tabulation in one corner of the sheet containing the pieces detailed. When very complicated drawings are dealt with, it may be accorded a separate sheet or sheets.

48. Working drawings may violate the rules of orthographic projection:—Projection is the theoretical side, working drawings are the practical application of theory.

Custom has santioned certain practices, more or less universal, for making drawings more explanatory with less labor, while there are innumerable short cuts, etc., adopted by different establishments, known only to the individuals having use for them. Some few of the general principles which may be followed will be here touched upon:

* (*a.*) "That in each separate view, whatever is shown at all should be represented in the most explanatory manner."

(*b.*) "That which is not explanatory in any one view may be omitted therefrom, if sufficiently defined in other views."

---

* McCord Mech. Draw. Part II, Page 3.

(*c.*) "The proper position of a cutting plane is that by which the most information can be clearly given."

(*d.*) "It is not necessary to show in section everything which might be divided by a cutting plane."

(*e.*) "Whatever lies beyond a cutting plane may be omitted when no necessary information would be conveyed by its representation."

The views necessary to show a subject do not follow the conventional ones of projection, for if one view is sufficient to tell the workman all the facts, more are superfluous; for example, one view of a bolt is all that is needed when the bolt is standard. The certainty that the workman could not make anything else from the drawings than the thing intended is the controlling condition.

When two pieces differ only in being rights and lefts, it is usually not necessary to draw but one of them, making an explanatory note on the drawing that two are wanted, one right and one left.

A section and an elevation are sometimes combined on the one view by superimposing the lines of the elevation over those of the section. This saves one view.

Lines coming very close when drawn to scale should be separated, that is, the scale exaggerated, or else one line left out.

In gears, a few teeth, perhaps only one, are drawn out in full; the remainder are indicated by dotted circles for their crowns and for their roots, the pitch circle being a dash and dot line, or the usual convention for center line, or it may even be made a solid line.

In sectioned views, continuity of material is not interfered with by the introduction of minor elements in the plane of the section. They are either left out or put in dotted.

Sometimes in the drawing of one part of an object, which it is particularly desired to show, there are other parts connected with it which may be rendered in dotted lines to help show the connection of them all.

And so illustrations may be multiplied, but it is not necessary to go farther. The different illustrations in the book will show some of the short cuts. Judgment and experience will open up others to the thoughtful draftsman, and he will even then occasionally find that there are opportunities for him to improve on past experience.

49. The development and arrangement of working drawings.

In beginning a set of working drawings of a subject which is entirely new in design, it is possible that the small features will be designed first, and the assembly drawings of parts, or of the whole, made afterwards, so there can be no rule for the order in which drawings are made.

Differences in manufacture and in the subject control the method of development and arrangement. Principles cannot be laid down applicable to all cases. Some one problem may be considered somewhat in detail, and will serve to show how it is done, and about the best illustration that can be taken is an academic exercise, a problem which would be given a student in drawing, that of making a set of working drawings of a model of a simple steam engine.

Determine the chief dimensions or size of the whole,

and choose such a scale which will properly present the assembly drawings, a plan, elevation and end view, upon one sheet without overcrowding the sheet. Next, take in turn the various parts, and make the necessary working drawings of each. It is best to take a survey of the size of the largest piece and of the smallest, and see of what size they can be conveniently made, using only one scale for all. Lay out the projections of the larger pieces first, and proceed toward the smaller, and so on.

Details of construction: Before any drawing is done, a list should be made of all the features needing to be detailed, and the number and kind of views required of each, a written list is preferable.

The arrangement of the sheets should next be decided upon. In practical, work different sized sheets are used for the different parts of the subject; frequently, the assembly views will be made upon a relatively large sized sheet, the main details on a second or medium sized sheet, and the smallest parts on small sheets. Small parts require, generally, finishing or machine work, and it is more convenient in the shop to handle the small sheets, mounted as they often are, on board, or some stiff backing, and varnished.

To get the best arrangement, a free hand sketch treatment should be used first, that is, to an approximate scale, sketch roughly, by very light lines, the space to be occupied by each and all of the views. This will permit of an adjustment in the arrangement, if it does not at first promise to be good.

Begin the careful drawing of the details first, leaving the assembly drawings until later, as the best interpretation and accuracy can be reached in working the assembly from the details.

Draw the views of the bed of the engine first. Do not begin at the top or bottom, and build steadily down or up until finished, but lay out the chief, or the over all sizes, then the next smaller, and so on, to the smallest parts last; also, drawing not one view at a time, but the several of the set; the same features recurring in the several views should be treated in them all, so that there should be harmony of parts.

If a view can be developed by projection from another, it is better to do so than to use the scale and lay it out independently, for it saves time. But the scale relation must be kept in mind and discrepancies noted.

The place on the sheet for the different views should be appropriate and follow a certain system. Large details should be put on a sheet by themselves, or else along the upper part of the sheet, or at the left hand side. The next smaller parts should be put below the first, or to the right, and so on, so that the smallest parts are shown along the bottom, or along the right hand edge of the sheet.

Related parts may be in projectively, related positions, provided the subject admits of it, or while not in projectively, related positions, they may be so intimately related on the drawing that the connection is apparent at a glance. For illustration: A connecting rod—showing the rod at the top—may have the straps to the left and right of the rod as if they had just been slipped off, their center lines coinciding with that of the rod; the brasses may be shown also to the right and left of the straps as if they had been removed by simply sliding along their center lines coinciding with that of their position in the straps; finally, the keys, and bolts, etc., may be put in the lower part of the sheet in any convenient place, arranged so that the left hand bolts, etc., belong at the left hand end of the rod, and those at the right, to the right hand end of the rod. This may be seen illustrated in diagram, Fig. 32.

Another logical and perhaps better arrangement is shown in diagram, in Fig. 33. Here the principle is followed of placing parts of a kind together, disregarding their exact position in the subject. The straps of both ends are put together at the left, but the upper one belongs to the left hand end of the rod, and the lower one to the right hand end of the rod, a certain convention of sequence which is quite common. Similarly, the brasses are placed at the right with, again, the left hand brass above and the right hand one below. The last mentioned plan of Fig. 33 is the best in general.

If the engine were complex, probably the connecting rods would be put on a sheet with the eccentric rods, or other long turned members, the brasses all together on a sheet by themselves, and the straps also. Even here, however, the rods for the high pressure (H.P.), low pressure (L.P.), and intermediate pressure (I.P.), would follow each other down or across the sheet in a certain sequence, which would be the same as that on the sheet of brasses and straps, etc. It is evident that such an arrangement would aid materially in reading the drawings and finding what is wanted.

If several parts of the engine are put on a sheet, the groups of drawings of them should be separated by a little

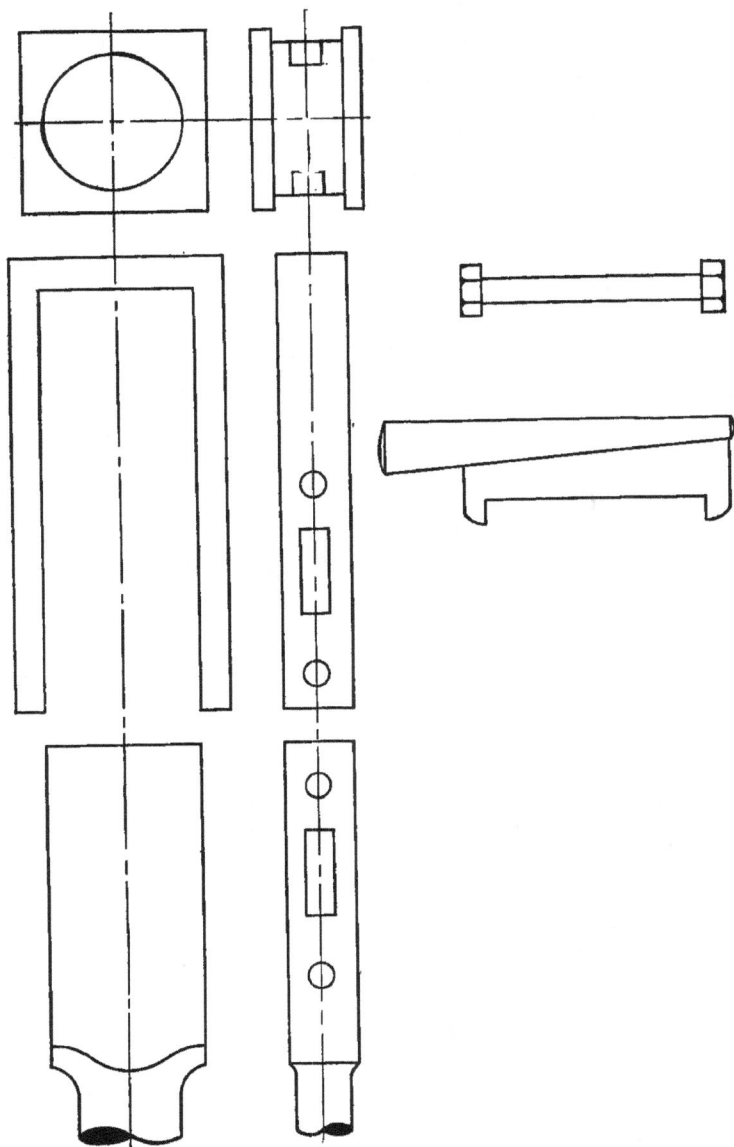

FIGURE No. 32.

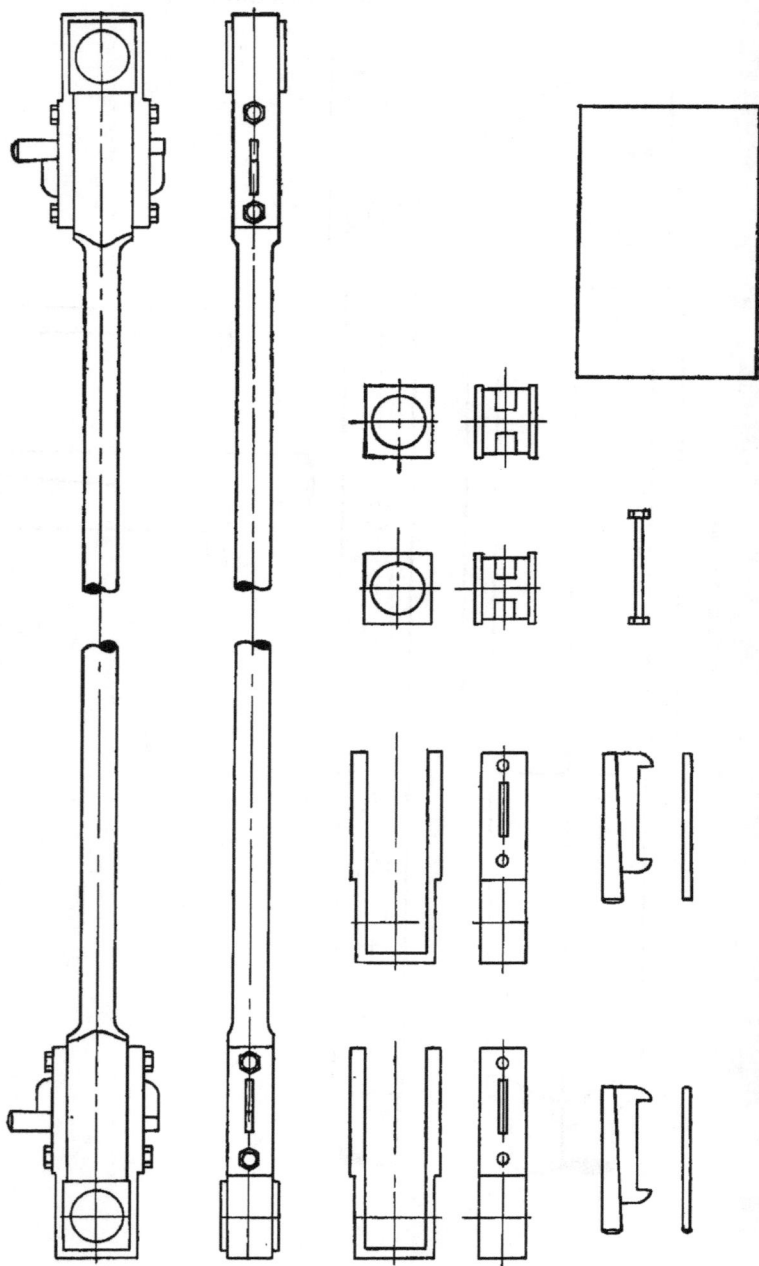

FIGURE NO. 88.

more space than the several views of a part, so that the
identity of the different things is not confused.

Parts which have to be made upon the same machine,
or by similar processes, are often collected on a separate
sheet by themselves; for example, there may be a sheet of
bolts alone, of screws, of forgings and of castings; but this
is done only where a large number of parts is wanted, and
where, moreover, processes of manufacture have become
somewhat systematized.

If it takes more than one sheet to make a set of
drawings, keep each sheet as far as possible self contained,
even though on some sheets there may be waste room.
The economy of space profits little, nor the even distri-
bution of views over the sheet unless it can be done without any sacrifice.

50. Drawing to scale is necessary where forms dealt
with are larger than the ordinary sized drawing paper will
take full size. It is no hindrance to the workman, because
he takes his sizes from those specified on the drawing and
is generally not permitted to use anything else.

There are a number of kinds of scales made, divided
broadly into civil engineer's and architect's or mechanical
engineer's and are either flat or triangular. *The civil
engineers'* scale is one in which the divisions of inches are
by even decimals, tenths, twentieths, thirtieths, etc. *The
flat scale* may contain two, four or eight scales, according
to the way in which its edges are divided, and is a convenient tool because of its flatness. *The triangular scale*
usually contains twelve different scales, and because of its
wide range is probably the favorite.

In engineering, when scale is mentioned, it means so many inches or fractions of an inch will stand for a foot of the actual thing drawn. To take a concrete case: On one face of the triangular scale, the edge is divided into 3-inch major divisions, identified by numbers on the flat surface, the edge is also divided into 1½ inch minor divisions, identified by numbers on the curved part of the scale. At the right a three inch space is divided into twelve major divisions to stand for inches and each of these again into eighths. At the left a 1½ inch space is similarly divided, except that each space standing for one inch is divided into quarters. Hence, by overlapping, we have two scales to an edge. To lay off a dimension with the three inch scale we read the even feet to the left of the zero mark and the inches or fractions to the right.

Points to be observed in the handling of the scale:—
It should never be used as a rule.

With a sharp and *round* pointed pencil, make a short straight stroke at the scale division, and perpendicular to the edge, about one thirty-second of an inch long.

Transfer measurements with the scale where possible, that is, indicate the size by scale measurement, then set the compass to the marks made if it is necessary to strike an arc of that radius.

Successive measurements should be laid off with one setting of the scale where possible.

The problems arising in the use of the architect's or mechanical engineer's scale group themselves under three heads.

(1.) To make a drawing a given fraction of the original in size.

(2.) Given the scale used to determine the fraction of size which the drawing is of the original.

(3.) Given the size of the space in which a drawing must be made to fit, to determine the scale to be used to get this reduction from the original.

(1.) To illustrate: Suppose it is desired to make a drawing one-quarter the size of the original; $\frac{1}{4} \times 12 = 3$; therefore 3 inches is the scale or size per foot to be used.

(2.) To illustrate: Suppose a drawing is made to a scale of one-quarter of an inch to the foot, then, as $\frac{1}{4} : 12$ so the drawing is to the original or $\frac{1}{48}$ the size.

(3.) To illustrate: Suppose a subject 3 feet long is to be reduced in drawing to $1\frac{1}{2}$ inches, the length being the determining dimensions, then $1\frac{1}{2} : 36 \times 12 = \frac{1}{2}$, or the scale is $\frac{1}{2}$ inch to the foot.

When problems do not come out as even as these, the nearest available scale is taken.*

There is a special and popular form of scale made for mechanical engineering work which differs in divisions from the ordinary scale. The inch and not the foot is made the unit for subdivisions. The scales shown are for half size, quarter size and eighth size. The half size, for example, has a half inch divided again into halves,

---

*A very common error arises from mixing up scale with fraction. A quarter scale drawing means a drawing made ¼ inch to one foot, while a ¼ size drawing means a drawing made 3 inches to a foot. In other words when we speak of a fractional size we do not mean that fraction as the scale but that fraction of 12 inches, the foot being the unit.

quarters and eighths, to represent those fractions respectively of an inch.

Sometimes a purely arbitrary and exact scale is required, and has to be constructed. It can be readily done by a method based upon the geometrical principle that lines parallel to one side of a triangle divide the adjacent sides in proportional parts.

The civil engineer's or decimated scale is mainly a scale for use when the reduction is relatively large so that from 10 to 100 feet will be represented by an inch, for it has divisions of 10ths, 20ths, 30ths, 40ths and 50ths of an inch. It can also be used in the same way as the mechanical engineer's scale. To illustrate: The twentieths scale can be used for $\frac{1}{4}$ inch to the foot, five divisions being equivalent to one foot, two and one-half to six inches, etc. The thirtieths can be used for six inches to the foot, five divisions then equalling one inch.

51. A section, or cut, through the whole, or part, of any subject is conventionally represented by covering the sectioned parts with evenly spaced parallel lines. They are not intended to represent or suggest a tinted surface, hence, to distinguish it as a sectioned surface the hatching lines, as they are called, are ruled diagonally of the edges of rectangular forms. Contiguous parts are section lined in opposite directions. In any position whatever of the contour lines of the sectioned surface, the section lines are generally made at an angle of 45° with them.

The weight of the section lines should not exceed those of the outline, they look rather better if made slightly lighter. Considerable judgment can be displayed in the adjustment of space and weight of section lines. Consulting practical drawings or the illustrations to this book will be helpful. The effect should never be coarse but pleasing, even, and not too obtrusive.

FIGURE NO. 84.

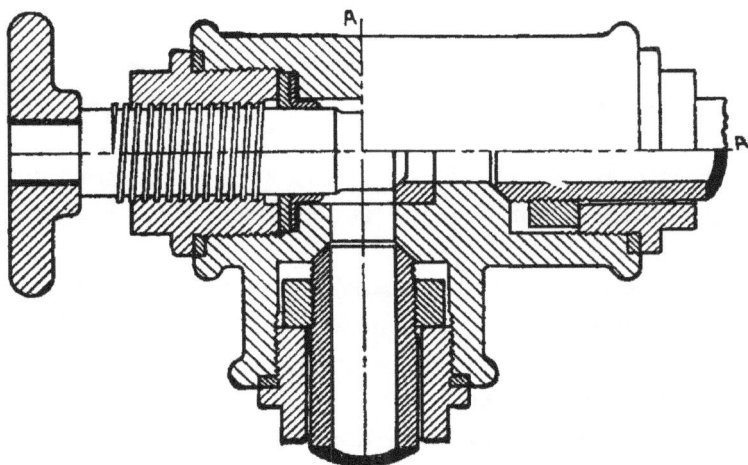

Fig. 34 is an example of sectioned surfaces. It can be seen that if narrow spaces are sectioned by relatively widely spaced lines, the effect is coarse. If on the other hand, the spacing is too narrow the effect will be that of a dark tone obscuring the outlines and any errors in spacing are more noticeable. Again, the larger the surface to be sectioned the wider can be the spacing of the lines. Long narrow pieces should be sectioned by heavy lines closely spaced and large areas by light lines widely spaced.

Draftsmen generally have a maximum and minimum of spacing of section lines that are not very far apart, nor do they use many different spacings because it is economical of effect if sectioning can be made more or less mechanical. One authority fittingly says: *"When a few lines have been done in section an unconscious rythmic action, as it were, is established, just as one beats time to slow or fast music without thinking, and the manipulation becomes mechanical. 'For this reason,' he goes on to say, 'a drawing to be inked should never be sectioned in pencil first, otherwise the result is likely to be as bad as if one were to write his name with a pencil and then try to go over the lines with ink.' "

Hence, it can be seen that this rythmic action comes more readily when the eye is accustomed to a very few different sizes.

Where more than two surfaces are contiguous to one another it is customary to use 30° or 60° lines to distinguish the third surface, but these angles are not resorted to unless it is necessary.

Sometimes contiguous surfaces are distinguished from one another by having the section lines fall uniformly short of the limiting lines of the surface by a small distance, say $\frac{1}{32}$ or $\frac{1}{16}$ of an inch as shown in Fig. 35. A section may be designated as a longitudinal or transverse section according to whether the plane of the section is parallel to the long axis of the subject or at right angles to it.

If the plane of a section is horizontal, it is called a *sectional plan*, if vertical, a *sectional elevation*, or in other directions it may be called simply a *sectional detail*. If a sectional plan, its place is above or below the elevation, according to the angle used in projection and it may take

---

* C. W. McCord, Mechanical Drawing, P. 6.

FIGURE 35.

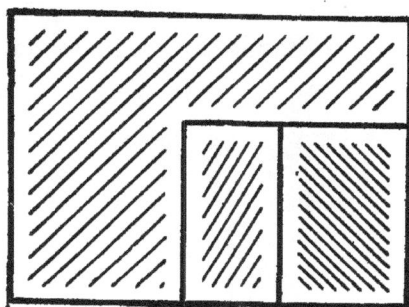

the place of a plan when it is not necessary to have the latter present. When it is, then the sectional plan may be placed either above or below the regular plan according to the angle of the projection, if there is not convenient room for this it may be placed to the right or left, orthogonally projected from the plan. Sometimes, according to circumstances, the section is placed independently of the regular projection views in any convenient place on the sheet. The importance of the section view and its value in giving data for construction determine the latter mentioned choice of positions.

To designate a sectional plan, a broken line, for which the convention is variable, is drawn across the elevation showing the position of the theoretical saw cut, and the ends of this line are lettered A—A or B—B, and so referred to in designating the sectional view (see Fig. 34). The broken line may consist simply of very heavy, short lines just entering upon, and on opposite sides of, the elevation, to call attention sharply to the place.

Sometimes forms are cut through which cannot be

accurately represented by a surface hatched with lines; the cross section of an I beam or a built up girder, as in Fig. 36. This is difficult because the surfaces cut through are

FIGURE NO. 36.

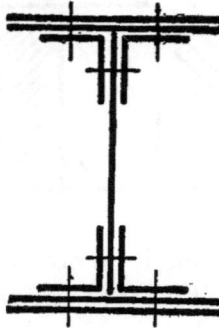

so narrow that it is impossible to represent them to scale by a double line, in fact the ordinary line on a drawing might itself be far too heavy. Scale in thickness of material then has to be sacrificed for effect in showing construction.

Forms are not always continuously cut through. Only a portion may be sectioned or the plane of the section may, for convenience or economy of views, be in reality two or more parallel planes cutting through two or more parts of the subject; very rarely are the planes made non-parallel, although such a condition is not prohibited by any written or unwritten laws. A very usual device is to represent a portion, say one-half, of an elevation as in section. This, if the subject be symmetrical upon a center line, will save the drawing of one view.

When the plane of a section moves from one place to another, as just explained, some conventional line, usually that for center lines, is used to divide the planes or the sectioned from the non-sectioned parts. The plane then terminated by this limiting line has no other limits. A solid limiting line is not used unless the material sectioned through actually ends at this place, and a new piece begins beyond it.

Where different pieces in the same sectional plane, are sectioned in a subject, considerable taste can be displayed in distinguishing parts one from another. Where the same piece reappears as cut through in the plane of the section it should be treated by section lines identical in weight and spacing so as to preserve, in other words, the continuity of material.

52. Standardized section lining has been attempted with but partial success by the adoption of certain weight and spacing of lines to be used as representing different materials, steel, wrought iron, cast iron, brass, etc. But, while a well known series has had wide publicity, that approved by the A. S. M. E., and in use by the Government drafting offices, still its use is not at all universal. Each drafting office has its own way of treating sectioning.

Fig. 37 shows the A. S. M. E. standards, together with the spacing and weight of line that are practical. Of course these are more or less governed by the area to be treated.

For wrought iron the groups of lines should be separated from each other simply by a little wider space than are the lines of the group, the same of steel.

The convention for glass is an attempt to illustrate the play of light one sees now and then upon looking into a

FIGURE NO. 87.

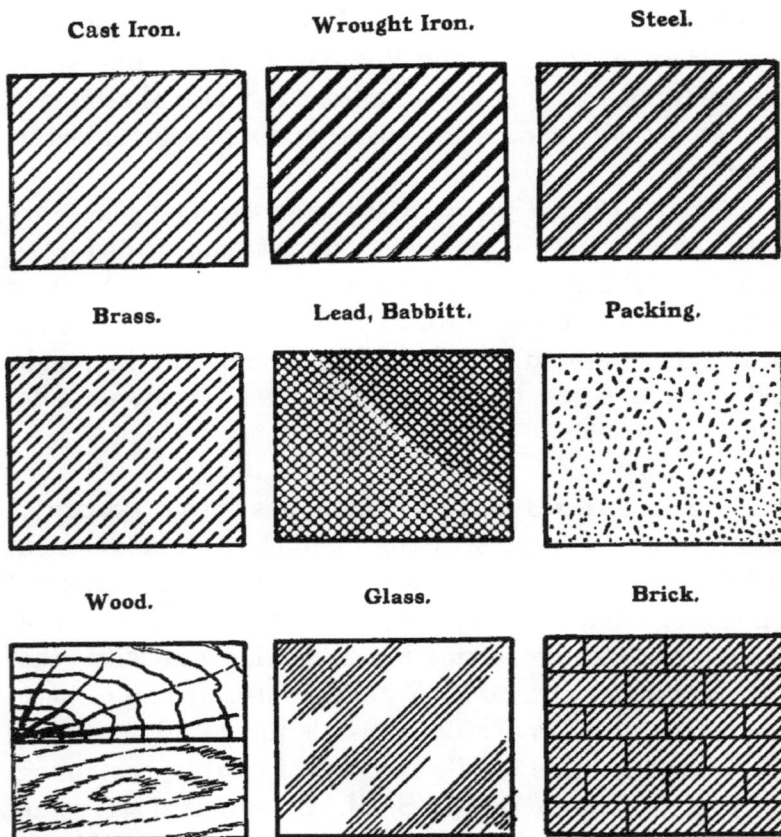

Cast Iron.   Wrought Iron.   Steel.

Brass.   Lead, Babbitt.   Packing.

Wood.   Glass.   Brick.

room from the outside through a window; the several masses of tinting, which by the way should be expressed by lines more closely spaced than in sectioning, are irregular in their contour shape, yet grouped together. The arrangement of this convention is after all a matter of taste of the draftsman.

Leather, sand or packing material are expressed by about the same convention. It is made with the writing pen; for packing, the strokes may take the shape of irregular lines like short threads.

In place of a conventional section lining, materials may be lettered W.I., C.I., S., etc., on the surfaces, to show what they are made of, or else a designating number may be used. The number may lie between limits which, by arbitrary agreement, stand for a certain material; for example, Nos. 1 to 200 for cast iron, 200 to 300 for wrought iron, etc.

53. Some practical points about sectioning.

Sectioned views are not used in working drawings unless it cannot be avoided or unless their use is economical of views, and consequently of time, and when not sacrificing clearness. They are used less as a vehicle for dimensions, also, than they are to show form, what is solid and what is hollow, what is continuous in the various materials. They are not desirable as a vehicle for dimensions because of the interference of the section lines with the dimension lines and figures. When dimensions have to be put across a dimensioned surface, space is left at least for the figures, sometimes for the dimension lines and the arrow heads. Aside from legibility, the presence of a dimension should be at once apparent through its prominence on the drawing.

A sectioned view should always show in full line that which lies within the plane of the section, all that is beyond this should also be shown in full line which is not covered by the surface sectioned. If it is desirable to show what is covered, it should be dotted as in any other

hidden construction.   Of course nothing should be shown
of that part supposed to have been removed when the
subject was cut through.

Section lining should be done directly in that medium
in which the drawing is to be left, and the spacing calcu-
lated with the unaided eye.   To do this well will require a
little practice; a few hints which follow may be found
helpful:—

>*Sketch the sectioned surfaces roughly free-hand*
>in pencil to show which are sectioned surfaces and
>which are not, and to show what direction of
>section lines to use.

>*Hold an arm of the triangle, when possible,* one
>edge of which is guiding the section lines, by the
>thumb and first or second fingers so that the
>triangle may be made to creep along the paper.

>*Use the edge of the triangle only for an approx-*
>*imate adjustment* of spacing, let the pen point give
>it more accurately.

Mechanical devices called section liners have been
invented for spacing lines evenly but they are not in
general use by draftsmen.   They fail because the pen
cannot be held so invariably the same way with respect to
the ruling edge that the section liner can be depended
upon to space automatically.

If a surface to be sectioned is broken up so that the
lines in all places cannot be drawn continuously across it,
as in a transverse section of a hollow cylinder, a line may
be followed along from its beginning to its end across the
subject before going on with the next one, or, a method to

be prefered is to draw a group of lines in one place and afterwards in another on the opposite side.

In the section view of features which wrap around one another like a valve with its stem, gland and body, it is desirable to begin with the innermost parts and develop outward from them.

Sometimes labor is saved, in hurried work, by not section lining entirely across relatively large surfaces but by sectioning only around the edges, stopping the section lines along imaginary lines parallel to the edges successively of the surface to be sectioned.

Long members of uniform cross section like I beams, etc., sometimes cannot be shown their true length in a drawing without reducing too much their transverse dimension. To overcome this they are assumed to be broken as shown in Fig. 38. The over-all dimension is

FIGURE NO. 88.

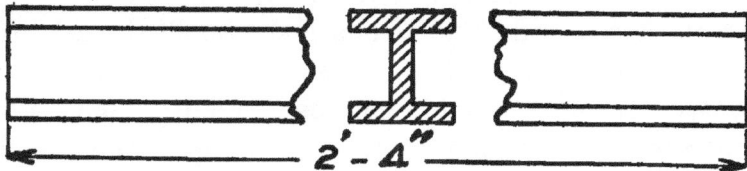

given and, to save a separate view, the exact shape of the cross section may be shown on one of the pieces, as if the plane of the section were turned through an angle of 90°.

54. The rules of orthographic projection are violated in sections, things are done which are not strictly projective or follow the theoretical saw cut. This is because clearness of construction is paramount.

For example, it is of no use to section longitudinally through a bolt or nut, the identity is not as readily distinguished. Bolts, washers, shafts, rods, or other solid pieces, having a relatively long axis proportional to their diameter, are never sectioned longitudinally.

The spoke of a wheel is not cut through longitudinally, if it should happen to come within the plane of a section. The rim and hub are the important features, the spoke is but a relatively narrow connecting link between the two. If sectioned, it is apt to give the impression of a wheel with a web or diaphragm connecting rim and hub.

It may be set down, as a general rule, that a rib, an arm of a pulley, or any comparatively thin piece should not be sectioned by a cutting plane which is parallel to its longest bounding surfaces.

The surface of a cut is not interrupted merely for the sake of showing a fastening or other minor feature. For example, if sectioning longitudinally through a cylinder and its head, and if the bolts fastening body and head should, some of them, happen to come within the plane of the section, they would not be sectioned, but either omitted or shown in dotted where they passed through the sectioned material and in full line elsewhere.

On the other hand, if we have a sectioned side view of any thing, having radially placed holes, and the plane of the section does not pass through one or more of them, nor is their arrangement clearly shown in another view, one hole should be cut through, and at its true radial distance from the center, to furnish information lacking elsewhere.

A key-way in a collar should not be sectioned longitudinally with the collar, but should be shown with a dotted line.

It is hardly worth while to multiply instances of violation of projection in sections; new cases are likely to arise continuously. Suffice it that features are not shown in section where no information would be gained thereby, no matter whether they come within the plane of the saw cut or not.

Sectional details are placed either near the part to which they are related or grouped together in any convenient place on the drawing, there is no rule governing. If placed near the principal form they are generally made projective with it, except that the sectional projection may be made on a supplementary plane not corresponding to the co-ordinate planes.

55. Dimensioning.

Dimensions are a most important part of a working drawing. If the scale on the drawing and the dimensions do not agree, the latter are assumed to be correct and govern the men in the shop. All that the workman needs to know must be put thereon, either in dimensions or footnotes, the latter being equally as important as the dimensions.

Sizes from a machine should always be taken with the foot rule and calipers, not with the scale. The rule should be put as close to the distance to be measured as possible; for accurate work, use the machinists' dividers, and apply them afterwards to the foot rule.

The machinists' dividers are used to ascertain sizes of flat surfaces, the calipers, inside and outside, are used to get the diameters of holes and cylindrical forms. After a diameter has been obtained with the inside calipers, one end should be set flush with the end of the foot rule. If convenient, place both against a flat surface. With the outside calipers, rest one of the arms against one of the end faces of the rule. These directions will facilitate the making of rapid measurements.

A knowledge of processes of manufacture will show that it is a waste of effort to measure everything to the thousandths of an inch, as some tight fits, of course, have to be. Rough castings cannot be measured to a sixteenth of an inch with accuracy. Work that is to be machined may or may not require to be very accurate. A tight fit may call for accuracy to a thousandth of an inch, another kind to a hundredth. Discrimination should be used in dimensioning. As an illustration of the application of approximate and exact measurements, take a line of sub-divided dimensions, which lie, on the one hand, between a finished surface, and the other the end of a casting. The last sub-divided dimension at the casting end should be omitted, and in its place an overall dimension should be given. The man who makes the casting will get the measurements he needs, while he who does the machine work and finishing will get those exact sizes he wants, and neither will, in any way, hamper the other.

All dimensions should be final working ones, no allowance should be made for shrinkage of castings, etc.

The distance between chief centers, and if the subject is symmetrical upon a center line, the distances also of these centers from the center line is always necessary.

The overall dimensions are needed in every case. The subdivided ones between depend upon the nature of the work which is to be done upon the particular piece, all the subdivisions may or may not be needed, they should be shown continuously on a line. It frequently happens that there are two or three main subdivisions, like the distances of the sides or ends of a piece from one or two important center lines, and in addition the distances of other smaller parts from these center lines. Where such exist we have three classes of dimensions. They ought generally to be figured up complete in each case to the overall, if for no other reason than that each may be a check upon the other. Aside from this, they may be needed in constructing the piece. The three sets should be close to one another, the overall the outermost, and the subdivided the innermost of the three. The relation between the three should be at once apparent.

The dimensions should be put on a design about as fast as the forms are constructed, because at any time a preceding dimension may be needed to find a subsequent one. And, furthermore, it is a check on what is needed for construction.

If the drawing is made from a model, as for study purposes, it is best to leave the dimensions until the last, as one thing to consider at a time leads to accuracy.

In constructing a drawing, and in setting off measurements and subdividing distances, use the scale where possible. Use the dividers or compasses only where it is unavoidable.

If the drawing is made from a model and the dimensioning left for the last, then the following system should prevail:

(*a*.) Start at the top of the original, and going down. note all those things requiring horizontal dimensions, and put the same on the views which will need them, or show them to best advantage.

(*b.*) Starting at the left, and proceeding toward the right, move similarly, recording the vertical dimensions.

(*c.*) Locate all radii, diameters and oblique dimensions.

(*d.*) Check everything.

These steps may again be subdivided into more. After locating some thing requiring a dimension, mark simply the place where it is to go on the drawing, by either sketching in the dimension lines and arrow heads free hand, or with the rule. After the place and the full number of all dimensions are located, then the foot rule should be applied, and the value of the several dimensions ascertained and recorded.

In inking the dimensions on a drawing made from the model, the same procedure should be gone through with except that the arrow heads should be put in first, and the dimension figures next. Nothing in the way of precaution, to preserve accuracy in dimensioning can be out of place. The inking of the limiting lines for dimension lines should precede the dimension lines themselves, for the same reason that the arrow heads should precede the figures. For, if the former are not done first, they are apt to be overlooked.

The practice is common in the better drafting rooms to give accurate dimensions in decimals and approximate in fractions. Structural steel drawings are dimensioned in decimals. When the dimensions are given in decimals, the inch or foot marks should be placed in front of the decimal to replace the whole number. Similarly, in case of dimensioning in feet and inches, and the inches are zero, or less than one, a zero mark with the inch sign over it should always be put in the inches' place, or a zero in front of the fraction.

Dimension lines are usually not made of any particular convention of line, because they are so varied in length. Generally, they are composed of several dashes when long, varying in length according to the distance to be dimensioned. Some drafting rooms use a solid line, but much lighter than any of the other lines of the drawing, and in rendering in ink where blue prints are to be made from it, they are made in red ink with a little black mixed with it to render it opaque, and make it show on the blue print as a very light blue line.

Dimensions should not be crowded between limits too narrow to receive them. The several ways of specifying linear dimensions, diameters and radii are shown in Fig. 39.

When a dimension is placed outside a form, limiting lines must be run to the form and perpendicular to the direction of the distance which is to be specified. They should be continuous lines, running just a trifle beyond the dimension line, and to a point just a trifle short of the outlines of the subject, as shown in the Fig. 39. The foot and inch marks may be as shown, either way, but the upper of the two is the best because there can be no misunderstanding of the symbol. The dash should always be put between the foot and inch figures, to prevent misunderstanding.

The dash marks when used for the feet and inches should be distinct, and easily distinguished from acci-

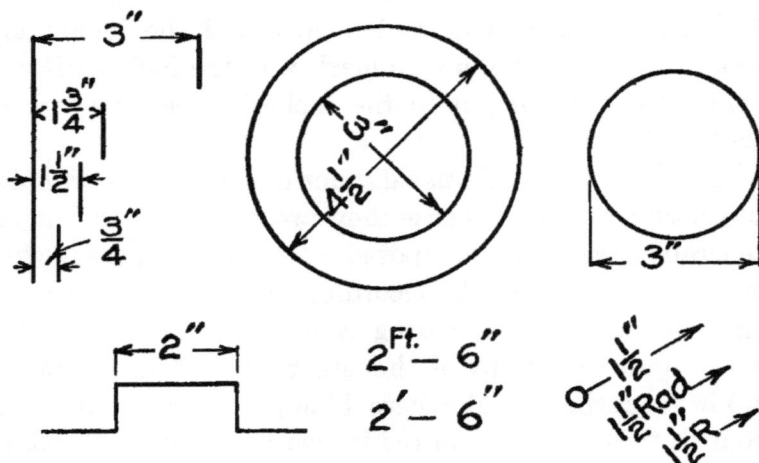

dental marks. They should go above and to the right of the figures, be about one-half the height of the figures, thick at the top and pointed at the bottom as the writing pen would naturally make them with the spreading of the nibs at the beginning of the stroke.

The arrow heads at the ends of dimension lines should be made with the writing pen, sharp and concave with the sides of the arrow at about an angle of 30° to one another. Care should be exercised to bring the point of the arrow to exactly the limit it is to accentuate.

Where leaders are taken from a point to a dimension, the line can either be ruled or made free-hand. The neatest appearance is attained when it is ruled.

The figures of a dimension should be printed, not written hurriedly. Figures that are from one to one and one-half times as wide as they are high are the best. Where fractions occur, the fraction line should be perpendicular to the line connecting the two figures, or in other words, horizontal when normally reading the figures. Recent practice puts the numerator exactly above the denominator and omits the fraction line.

The fraction figures may be made as large as those of the whole number, but also can equally appropriately be made slightly less. A good working rule is to make them just a trifle smaller than the whole numbers.

Notes should be made concerning materials or finish, written on the drawing in off-hand lettering or they may be included in the bill of materials. If the latter, the note will state the material used, how each part is to be finished, and the number of pieces required. Special directions pertaining to making, painting, shipping, etc., may even be given in notes; also sometimes, notes pertaining to erection are added, like "These rivets are to be field driven."

To summarize briefly:—

Those dimensions should be put on which are needed in the various processes through which pieces are to be put in making, should be clear and not crowded, and should furthermore:—

Read from the bottom and right hand sides.

Arrow heads should be small, neat, sharp pointed, clear and with small angle between the sides.

No dimensions should be put on center lines.

No lines should be drawn through dimensions.

All dimension figures should be of the same size.

Dimension figures should be in the center of the dimension line, where possible.

All figures should be of the same size.

All measurements, in general, should be made from center lines or from finished surfaces.

In cross sections, leave a clear space unsectioned
for the dimensions.

The most appropriate dimensions should be on
each view.

Similar parts should be dimensioned in like places.

Dimensions should not, in general, be repeated.

Dimensioning of certain features:—

A tapped hole is one bored to receive a screw and is
dimensioned for the required diameter of bolt or screw, as
for a $1\frac{3}{4}''$ screw. If cast, the hole is specified as cored for
a sufficiently smaller size to permit of its being tapped out
to the required size. If a hole is tapped for a bolt, the
drawing may or may not show two concentric circles to
stand for the base and the crown of the thread. If only
one is shown, it will be the circle which specifies the diameter of the screw, namely, the crown of the screw thread.

Bolts for a tapped hole will, if standard, only require
the size or diameter which may be specified as a dimension, or by a note to one side, the length under the head
and the length that is threaded, the latter specified generally as a distance from the end of the bolt. If the bolt
is not standard, it will require, in addition to the above
dimensions, the height and diameter of the head and
number of threads per inch. If the end of a screw is
rounded, the dimensions of the overall and the length that
is threaded should be given to the corner and not to the
extreme end.

When bolt holes are spaced equally around a center
upon a disc or a cylinder head, for example, a circle passing through all their centers, together with radial lines,

also passing through the centers, constitute center lines for these forms, and the radius of this bolt circle is given, and also the number of the bolts, or the angular distance between them.

Angles, if specified, may be given in degrees, or by co-ordinates or by tangents, depending upon circumstances. If in degrees, an arc is struck from the vertex of the angle as center, the dimension line constituting this arc; if by co-ordinates, any two distances at right angles to each other are used, measuring from each other and the vertex of the angle; if by tangents, it is the length of a perpendicular measured from a base, which is one side of the angle, of length one, measuring from the vertex of the angle. Measurement by co-ordinates is of particular use to the pattern maker.

Always dimension filleted or rounded corners, where an adjoining finished surface does not prevent and give the radius of the fillet, as its size may be important in adding strength to the angle.

The dimensions concerning a number of rivets or small bolt holes in a right line, should be written between the arrowheads of an overall. It should contain the number, size, distance apart, as well as the distance of centers overall. The distance of the end holes to the end of the piece should regularly be given.

Where a taper is required, it should be dimensioned with the taper per foot of length. Occasionally, it is specified by giving the dimension at each end of the taper; where this is done the approximate taper ought also to be given.

The dimensions of boards and iron plates should be

specified in the order of width, thickness and length, for example a board $14''x\frac{1}{4}''x8'-0''$. The grain will be parallel to the $8'-0''$ edge.

56. Notes for kind of finish are required on some drawings, as file finish, grind finish, etc.

For plane finish, an *f* is put across the edge view of the surface to be so treated. A note may be added, when the finish is to be of a different kind. It is necessary to designate it so that extra material may be allowed in the casting.

A few of the kinds of finish are as follows:—

Polish:—Smooth and glossy surface without accurate adherence to dimensions.

Finish and polish: — Smooth and glossy surface without sacrifice of accuracy of work.

Spot face:—Making the spot true.

Finish and scrape:—Shape to dimensions by cutting tool and afterwards scrape the surface.

Finish and grind:—Shape to dimensions with cutting tool and grind.

Trim: — Shaping by any convenient means as chipping, filing, grinding, etc.

57. Drawings are marked with symbols in the lower right hand corner and also in the upper left hand corner, upside down, so as to be read whichever way the drawing happens to lie. The character of the mark depends upon the system of filing used. In some cases, a certain class of machines has numbers between certain limits and a letter arbitrarily designates the size of the sheet, as D—188.

58. **The checking of drawings** should be done by methodical steps; a bird's eye view is not a sufficient check. The drawing should be checked through approximately the duplicate of steps by which it was originally made, and each step should be carried carefully throughout before the next step is undertaken, as follows:—

(1.) Identify every piece of a subject to see if all are fully shown, and also the requisite views of each.

(2.) Note lines of various views for completeness and correctness.

(3.) See that all dimensions are given that are needed, also working notes.

(4.) Scale every dimension to see if it is .correct, putting a check mark alongside of each as checked.

(5.) See if they correspond with each other in different parts, in the assembly and the details.

(6.) See if arrow heads are all shown.

(7.) See if dimensions are well placed.

(8.) See if accents for feet and inches are all correct, and fractions or decimals plain.

(9.) See if center lines are all in and correctly shown.

(10.) Finally check for supplementary notes and directions, including bill of materials, if there is any.

Checking is often done by other than the draftsman. The final checker, perhaps, may be a man who does nothing else, and who is held responsible for the results after the drawing leaves the drafting room.

59. Conventions in common use on working drawings:—

There are a number of conventional methods of representing forms and certain constructions, which recur very frequently in working drawings.

Different establishments have their particular conventions, used in diagraming, etc., among these conventions, a characteristic group is those used in electrical work. Fig. 40 shows a few of these.

FIGURE NO. 40.

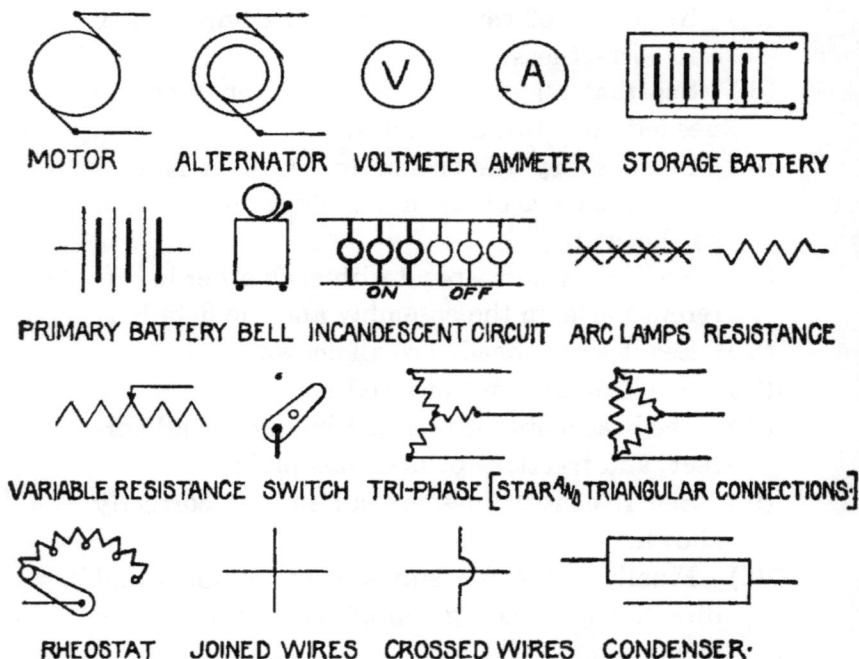

MOTOR   ALTERNATOR   VOLTMETER AMMETER   STORAGE BATTERY

PRIMARY BATTERY BELL INCANDESCENT CIRCUIT ARC LAMPS RESISTANCE

VARIABLE RESISTANCE SWITCH TRI-PHASE [STAR AND TRIANGULAR CONNECTIONS]

RHEOSTAT   JOINED WIRES   CROSSED WIRES   CONDENSER·

When a long thin member like an angle, or T bar, or I bar, etc., is broken for any reason, the approximate shape of the section is shown on the end. Sometimes the

accurate shape is given so that one view may do for two.
See Fig. 41.

FIGURE NO. 41.

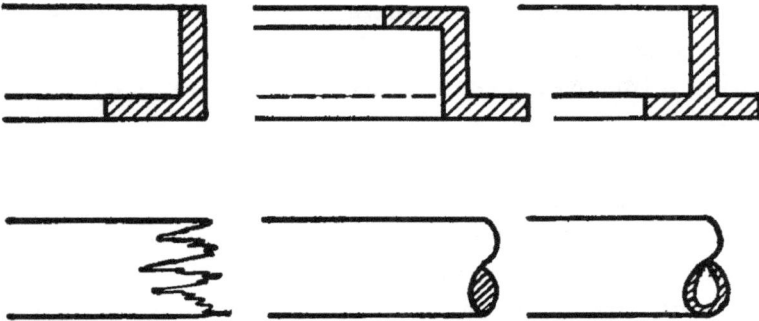

A *round shaft or rod is broken*, as shown, and if
hollow, the approximate thickness of the metal is indicated. The curve of the break may be put in with the
curved rule, but also may be done free-hand. Wood is
shown by representing the splintering that is apt to
accompany a break.

Now and then colors in very pale tints are used to
represent various materials of construction. They ought
only to be put on drawings which are stretched to the
board. The following are the most important of these:

Cast iron, *payne's grey.*

Wrought iron, *prussian blue.*

Steel, *prussian blue with tinge of carmine.*

Brass outside, *gamboge.*

Brass in section, *carmine.*

Grained or knotted wood, *burnt sienna.*

Earth, *burnt umber.*

Brick, *light or venetian red.*

Masonry, *wash of india ink with tinge of blue.*

More natural effect can be given to other materials like wood, water, etc., according to the artistic skill of the draftsman.

In some elaborate drawings, a nice effect is obtained by shading, where possible, with the india ink lines and putting a tint over these of the convention for the material shown.

Since the ability to put on a good flat tint is a necessary accomplishment of the civil engineer, the following extracts upon the subject are quoted from F. N. Willson's "Theoretrical and Practical Graphics."

"The surface to be tinted should not be abraded by sponge, knife or rubber."

"The liquid employed for tinting must be free from sediment, or if the latter is present, it must be allowed to settle and the brush dipped only in the clear portion at the top. Tints may, therefore, best be mixed in an artist's water-glass, rather than in a shallow receptacle."

"Tints are best prepared from the india ink in cakes, and from other water colors in the pans. The size of the brush should bear some relation to that of the surface to be tinted."

"Since tinting and shading can be successfully done, after a little practice, with only penciled limits, there is but little excuse for inking the boundaries; but if for the sake of definiteness, the outlines are inked at all, it should be before the tinting, and in the finest of lines, preferably of 'water proof ink,' although any ink will do, provided a soft sponge and plenty of clean water are applied to remove any excess that will 'run.' The sponge is also to be the main reliance of the draftsman, for the correction of errors in brush work; the water, however, and not the friction, to be the active agent. An entire tint may be removed in this way if it seems desirable."

"When beginning work incline the board at a small angle, so that the tint will flow down after the brush. For a flat tint, start at the upper outline of the surface to be covered, and, with the brush full, yet not so as to prevent its coming to a good point, pass lightly along from left to right, and on the return carry the tint down a little further, making short, quick strokes, with the brush held almost perpendicularly to the paper. Advance the tint as evenly as possible along a horizontal line; work quickly between outlines, but more slowly along outlines, as one should never overrun the latter, and then resort to 'trimming' to conceal lack of skill. It is possible for any one, with care and practice, to tint to, yet not over, boundaries."

"The advancing edge of the tint must not be allowed to dry until the lower boundary is reached."

"No portion of the paper, however small, should be missed as the tint advances, as the work is likely to be spoiled by retouching."

"Should any excess of tint be found along the lower edge of the figure, it should be absorbed by the brush, after first removing the latter's surplus by means of blotting paper."

"To get a dark effect, several medium tints laid on in succession, each one drying before the next is applied, give better results than one dark one."

"A tint will spread much more evenly on a large surface, if the paper is slightly dampened with clean water. As the tint will follow the water, the latter should be limited exactly to the intended outlines of the final tint."

60. Tracings:

Usually blue prints are made from drawings to use in shops, etc. Tracings, or tracing cloth are the best for this purpose. The drawings may either be made on manila or detail paper, as it is called, or they may be made directly upon the rough side of the tracing cloth.

A few directions are necessary upon the handling of the tracing cloth. The penciling, as well as ordinary dirt and soil, can be cleaned off by rubbing with gasoline, ether, benzine, or any highly volatile substance. Before inking on the cloth, a little chalk should be rubbed over the surface with a rag, for there is more or less grease apt to be present, and it interferes with the drawing of lines in ink. If the fibres should get at all torn or injured, they may be repaired, partly, by rubbing with soapstone or hard beeswax. The soapstone comes convenient in the form of the soapstone pencil.

Special care is necessary in inking on cloth, particularly if the smooth side is used. The lines made by the pen are apt to be thicker by spreading, and consequently blots are easy to make. The precautions to be observed are to carry less ink in the pen, than if working on paper, and to be sure that lines are dry before working up against them. Speed in crossing a line, or working from and to lines, is necessary. Mistakes are not so easy to correct.

61. The V and square threaded screws are based upon the curve know as a helix. If a point moves around the surface of a circular cylinder at a uniform rate, and at the

same time moves at a uniform rate in the direction of the
axis of the cylinder, it will generate the helix. From
co-ordinate geometry, a curve plotted between co-ordinates
which have a directly proportional relation to one another,
will be a straight line. Hence, the helix may be defined
as the shortest line which can be drawn upon a circular
cylinder between two points that lie neither upon the same
right section or upon the same right line element.

62. To study the helix in projection, draw a cylinder in
the third angle (as Fig. 42) with axis parallel to the vertical
plane. Assume a point to be at O, and to move around
the cylinder in the direction of the arrow through equal
distances, 1, 2, 3, etc. Let it move also up the cylinder
through any given distance, until, after it has completed
one revolution of the cylinder, it reaches a position, p',
directly above o'. The distance, $o'p'$, is known as the
pitch of the curve. As both motions are uniform, the
point will travel to 1, which is one-twelfth of the circumferential distance in the same time that it travels one-
twelfth of the distance of $o'p'$ towards p', and to 2, which
is one-sixth of the circumferential length, as it goes one-
sixth of the distance, $o'p'$, towards p' and so on. Hence,
to plot the curve, divide the circumference of the plan into
any convenient number of equal parts, and the pitch into
the same number of equal parts. By noting the points of
intersection of the perpendiculars to the ground line,
through the divisions of the circumference and parallels to
the ground line, through the corresponding divisions of the
pitch, points of the curve may be found.

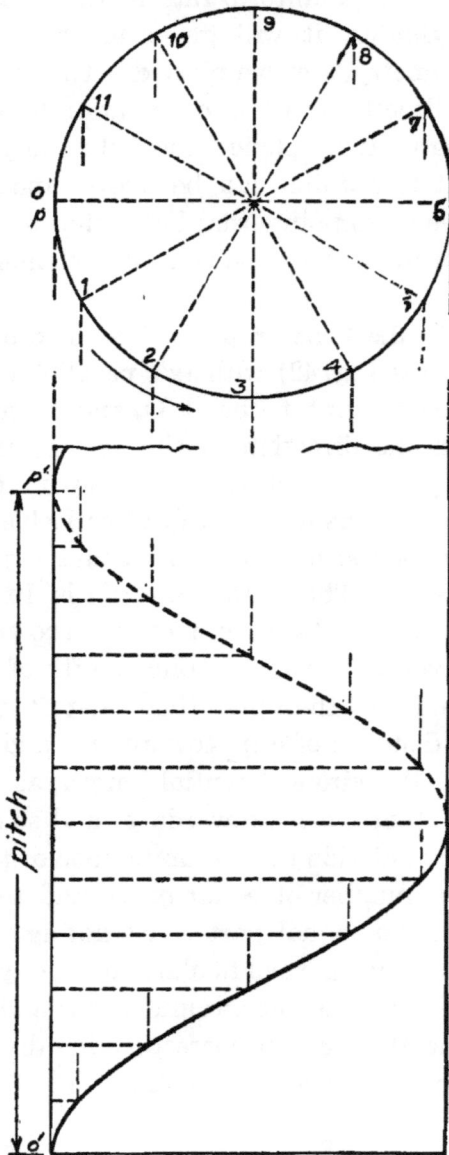

FIGURE No. 42.

63. Certain peculiarities of the curve deserve notice: (*a.*) It is tangent to the contour elements of the cylinder at points, *o'* and 6. (*b.*) It changes curvature at point, 3, midway of the contour elements, The tangent to the curve at 3 shows the angular pitch, which is the ratio of the linear pitch to the circumferential distance. (*c.*) The curve is sharpest at *o'* and gradually grows straighter until at 3 it reaches a straight line for a very short distance. (*d.*) It is symmetrical in parts with respect to the axis of the cylinder, and to lines perpendicular to the axis so that the curve from *o'* to 3 is a unit which is repeated throughout the path of the point.

If the pitch is lessened, the curve, at the contour elements of the cylinder, grows sharper, and at the middle of the cylinder straighter, approaching throughout straight lines oblique to the axis of the cylinder. When in the ordinary screw the pitch is exceedingly small, relative to the diameter of the screw, it is next to impossible to draw the curve correctly. The pitch in the screw means the distance between the turns of the thread measured axially of the screw.

64. In the V threaded screw three curves appear, one theoretically wound around a larger cylinder at the crown of the thread, and two wound around a smaller cylinder, at the base, or root of the thread. In the square threaded screw, four curves appear, two at the crown and two at the root. These two threads are shown in Figs. 43 and 44.

The contour lines of the V thread do not meet at the crown of the thread in sharp angles, but each is tangent to the curve of the crown. Again, these lines do not meet at

FIGURE No. 48.

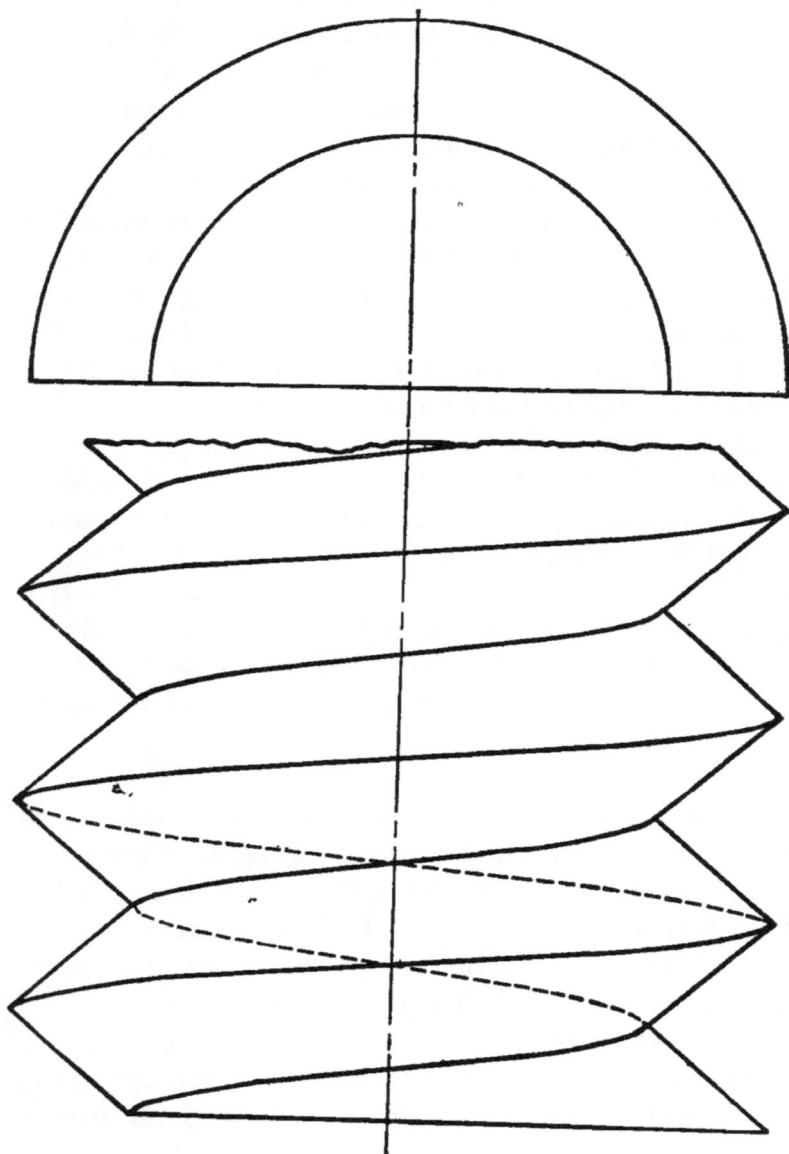

FIGURE No. 44.

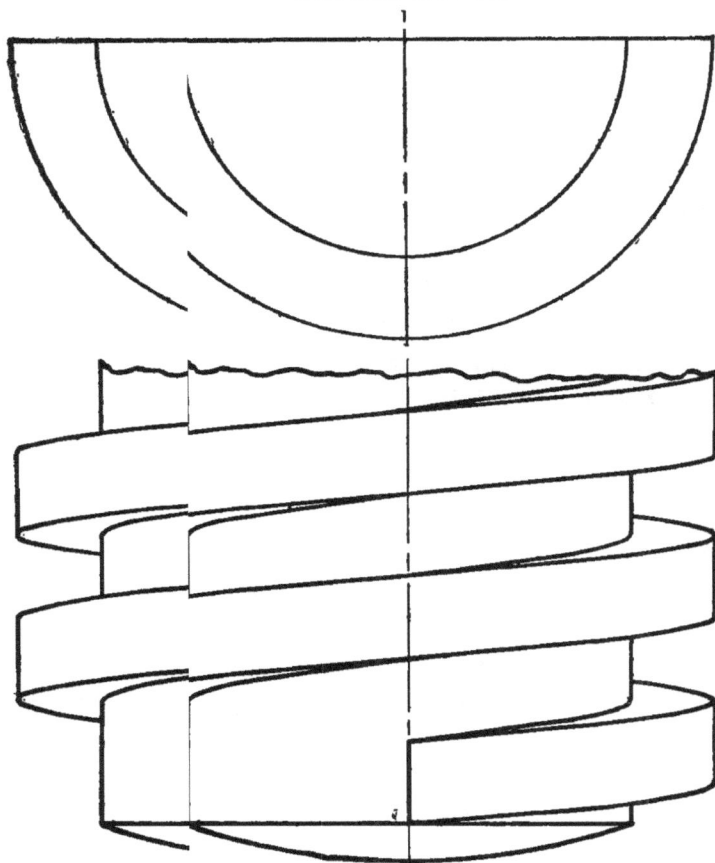

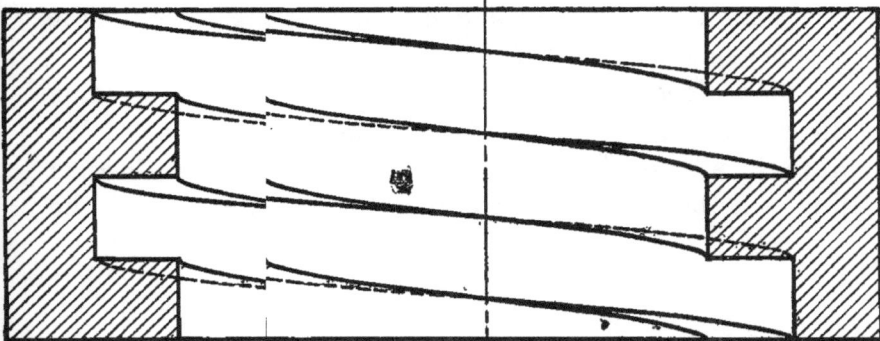

a point which lies on the cylinder upon which the root curve is theoretically wound, but they are tangent to the root curve, and cross one another a little outside of it. These facts are neglected in any practical drawing of the thread, but should be comprehended. They are, unfortunately, too often incorrectly shown in careful drawings of the thread and in well known text books.

The diameter of a screw, as dimensioned, always stands for the diameter of the crown line of the thread; this does not indicate the strength of the screw. The strength is determined by the diameter at the root curve of the thread, being the minimum cross section of the material of which the screw is made.

65.   Threads are shown conventionally unless in exceedingly rare cases.   The first change from the accurate thread is to make the curves straight (as shown at *a* and *b*) in Fig. 45.   The next change, in the case of the V thread, is to omit the saw tooth edge, leaving just the longer and shorter lines, as shown at *e* and *f*, making the limits of the screw a cylinder.   Another convention is shown also, one which suggests, in a way, roundness with a sacrifice of the screw characteristic as at *g*.   Square threaded screws are relatively rare, and there is no particular convention in use to represent them beyond making one thread of full lines and dotted limiting lines for the others as at *c*.   Sometimes this is done in the V thread.

The lines of the thread have a slight inclination upward toward the right; the direction of the slant of the lines, in all positions of the screw, can be ascertained in this way by looking in the direction of the axis of the

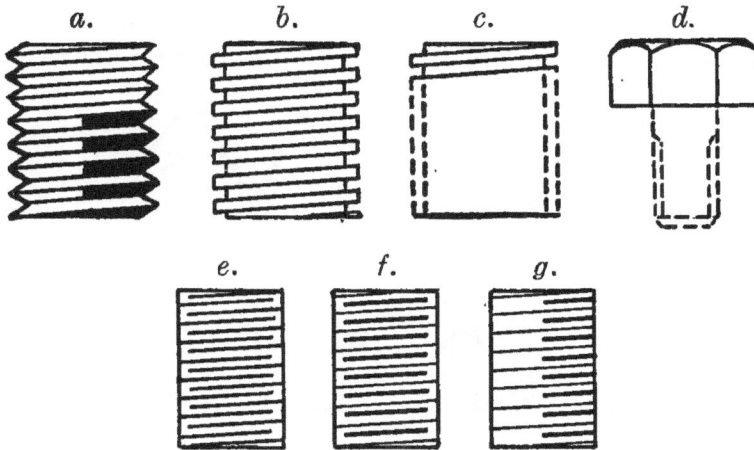

a.   *b.*   *c.*   *d.*

e.   *f.*   *g.*

screw. For left handed threads, which are rare, the slant
is upwards toward the left. The tap for a screw (see Fig.
44), if shown in section, will have its lines the reverse of
those in the screw, because it is the duplicate of the curves
on the rear half of the screw, which in the latter are not
seen.

In ordinary drawing, of course the pitch, as conventionally treated in Fig. 45, is not measured but estimated.

66. The Whitworth standard has an angle of 55°
between the sides of the thread. The crown and root of
the thread are both rounded off. The amount taken from
the crown, and that added to the root of the thread, being
equal to one-sixth of the total depth of the thread. Let
$D' =$ diameter at the bottom of the thread, $D =$ outside
diameter of the thread, and $N =$ number of threads per
inch; then $D = D' - \dfrac{1.28065}{N}$ (see Fig 46 a).

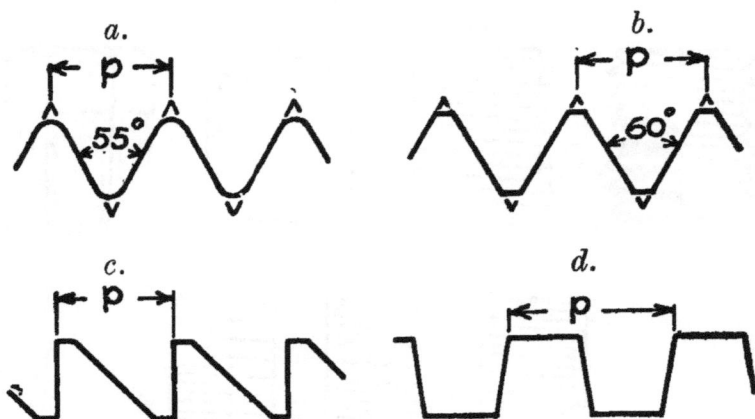

67. **The U. S. standard** proportions, devised by the late William Sellers of Philadelphia, has an angle of 60° between the sides of the thread. The crown and root are cut off flat; the amount added at the root being equal to that taken from the crown; the depth of the flattened face being equal to one-eighth of the depth of the thread. Using the notation as given above, $D = D' - \dfrac{1.299}{N}$ (see Fig. 46 b).

68. **There are other forms of thread** occasionally used besides the V and square thread. The buttress thread, for example, is shown in Fig. 46, at c. It is used where the screw is a transmitter of power or to resist force in one direction. A screw to transmit motion may have a short or steep pitch according to the speed of the screw. If the speed is rather great, and yet it is desired to keep the screw strong, or even for the latter reason alone, it may have a double thread or a triple thread. Thus, for one revolution of the screw, it will travel axially two times the pitch, or three times the pitch, etc. An illustration of a screw with steep pitch, to give relatively large axial motion, is shown in the thread on the spindle of some valves.   There is even a certain point which can be reached in the pitch of a screw thread, that longitudinal pressure of the screw bearing against the thread will turn it.   This is illustrated in the self-acting screw drivers.

A truncated V thread is one in which the crown and root have been very materially flattened. It is a kind used in the spindles of some valves, as before mentioned A form of truncated V thread, known as the Powell thread, is shown in Fig. 46, at *d*.

69. **Bolt heads and nuts are almost universally made of the hexagonal, or square form,** so as to be convenient to grip with a wrench. The hexagonal form is preferable, because in cramped places the hold of the wrench can be changed after turning through an angle of 60° (see Fig. 48.

The sizes of the heads, both hexagonal and square, are universally standard for those in common use. Bolts are either cast, cut or drop forged. The sizes are determined as follows: Let d = the diameter of the bolt, D the diameter of the head, then the formula is $D = 1\frac{1}{2}d + \frac{1}{8}''$. D is the true diameter of the hexagonal and square forms, namely, the perpendicular distance between the middle of opposite faces. It is taken this way so that the wrench for either will be the same.

The heads are not left in the prismatic form, exactly, but are chamfered. The under side of the head is left flat. In nuts, both hexagonal faces are sometimes chamfered.

A nut, that is to tighten down on a flat surface, will work better into place without cutting the material, if its corners are rounded. To get the chamfer, the bolt is put in the lathe, and the cutting tool set at 45° to the axis of the bolt, and the head cut until all the edges are removed. The surface of the cut is a cone; the intersections of the plane surfaces of the head with this cone give hyperbolas, very short arcs however, differing so slightly from the circular that never in practice would they be drawn accurately. In finished bolt heads, the dimensions are somewhat less than the formula, being $D = 1\frac{1}{2}d + \frac{1}{16}$. The height of a bolt head and that of the nut vary with the diameter of the bolt, but are generally drawn equal to it, Fig. 47 shows a table of useful facts concerning bolts, etc.

Fig. 48 shows the customary method of drawing the hexagonal bolt head. The lines are self explanatory. There are two forms of hexagonal heads, the rounded and the square, The former is shown in the figure and is a cone cut by seven planes, six for the sides and one for the top of the head.

Bolt heads and nuts should always be shown alike in position of a drawing. The principal view, if of the side of a hexagonal bolt, should show three faces, of a square bolt head, two. If only one view of a bolt is shown, it should be that showing three faces of the hexagonal or two of the square headed.

If a lock nut is used its height may be taken as one-half the diameter of the bolt.

In dimensioning standard bolts, it is customary to give only the diameter of the bolt, the length under the head and the length which is threaded, measuring from the tip

FIGURE NO. 11.

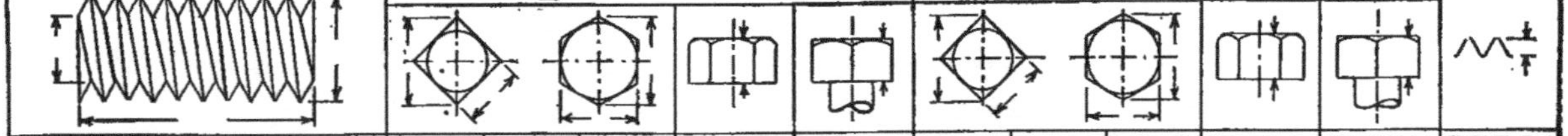

DIAM	THD'S PER INCH	DIAM ROOT TH'D	DIAM TAP DRILL	ROUGH LONG DIAM OF SQ	ROUGH SHORT DIAM BOTH	ROUGH LONG DIAM HEX	ROUGH THICK-NESS NUTS	ROUGH THICK-NESS HEADS	FINISHED LONG DIAM OF SQ	FINISHED SHORT DIAM BOTH	FINISHED LONG DIAM HEX	FINISHED THICK-NESS NUTS	FINISHED THICK-NESS HEADS	FINISHED DEPTH OF TH'DS
1/4	20	0·185	3/16	23/32	1/2	19/32	1/4		21/32	7/16	17/32	3/16	3/16	1/32
5/16	18	0·240	1/4	27/32	19/32	11/16	5/16	19/64	25/32	17/32	5/8	1/4	15/64	1/32
3/8	16	0·294	5/16	63/64	11/16	51/64	3/8	11/32	59/64	5/8	47/64	3/8	9/32	3/64
7/16	14	0·344	23/64	1-7/64	25/32	29/32	7/16	25/64	1-3/64	23/32	27/32	7/16	21/64	3/64
1/2	13	0·400	13/32	1-15/64	15/16	1	1/2	31/64	1-1/64	3/4	5/16	1/2	3/8	1/16
9/16	12	0·454	15/32	1-1/2	1	1-1/8		5/8	1-1/16	7/8	1-1/16	9/16	27/64	1/16
5/8	11	0·507	17/32	1-49/64	1-1/16	1-7/32	9/16	17/32	1-45/64	1	1-5/32	5/8	15/32	1/16
3/4	10	0·620	5/8	2-3/32	1-1/4	1-21/32	3/4	5/8	2-15/64	1-3/8	1-3/8	3/4	9/16	5/64
7/8	9	0·731	3/4	2-19/64	1-5/16	2-5/32		25/32	2-49/64	1-5/16	1-19/32	13/16	3/4	3/32
1	8	0·837	27/32	2-53/64	2	2-5/16	1	29/32	3-19/64	1-1/4	2-3/32	15/16	23/32	3/32
1-1/8	7	0·940	15/16	3-3/32	2-3/16	2-17/32	1-1/8		3-19/64	1-5/16	2-1/4	1-1/32	27/32	7/64
1-1/4	7	1·065	1-3/32	3-23/64	2-3/8	2-23/32	1-3/8	3/32	3-19/64	1-5/16	2-1/16	1-3/16	15/16	7/64
1-3/8	6	1·160	1-3/16	3-5/8	2-9/16	2-31/32	1-1/2	3/16	3-9/16	2-1/8	2-9/32	1-5/16	1-1/16	9/64
1-1/2	6	1·284	1-1/4	3-57/64	2-3/4	3-3/16	1-3/4	3/16	3-53/64	2-1/4	2-29/32	1-7/16	1-3/16	9/64
1-5/8	5-1/2	1·389	1-5/16	4-5/32	2-15/16	3-13/32	1-7/8	15/32	4-3/32	2-11/16	3-1/8	1-9/16	1-13/32	5/32
1-3/4	5	1·491	1-3/8	4-27/64	3-1/8	3-5/8	2	9/16	4-23/64	3-1/16	3-11/32	1-11/16	1-1/2	5/32

FIGURE No. 48.

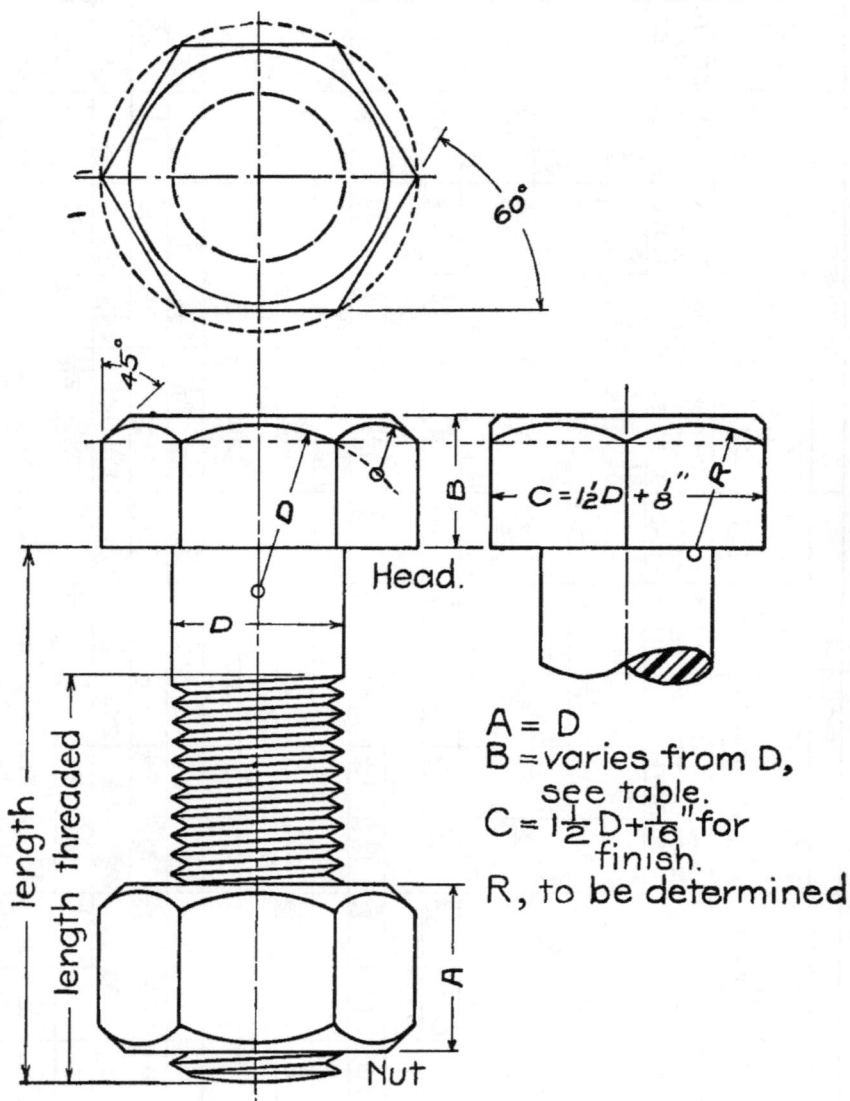

Head.

$C = 1\frac{1}{2}D + \frac{1}{8}''$

Nut

$A = D$
$B = $ varies from D, see table.
$C = 1\frac{1}{2}D + \frac{1}{16}''$ for finish.
R, to be determined

of the bolt, unless the bolt has a flat or countersunk form of head in which the 'over all' is the required dimension of length.

70. Bolts and screws are variously named according to the functions they perform.

An ordinary bolt and nut is generally used to fasten two or more pieces together, may be of any length and threaded any length, It may be rough or finished. It may or may not have two washers one under the nut and one under the head to form good bearing surfaces. Finally, it may have any one of a variety of shaped heads.

A machine bolt is one having the characteristics of an ordinary bolt as well as others, except that it is finished all over and the head and nut are consequently smaller in dimensions than those in rough bolts.

The various forms of heads to bolts, excluding the hexagonal or square forms, are shown in Fig. 49, with the standard proportions in common use.

A tap bolt is used on the rougher classes of machine work usually, to fasten two or more pieces together without the aid of a nut, namely, by tapping into one of the pieces. It should not go down to the bottom of the tapped hole. It is usually threaded throughout its length, although this is not invariable.

A cap bolt or cap screw is a form of tap bolt, so named from its function in holding down caps or covers. It is usually threaded for three-quarters of its length, in the smaller sizes.

A stud bolt is one which is screwed into a tapped hole in one piece, and after another piece has been slipped over

FIGURE NO. 49.

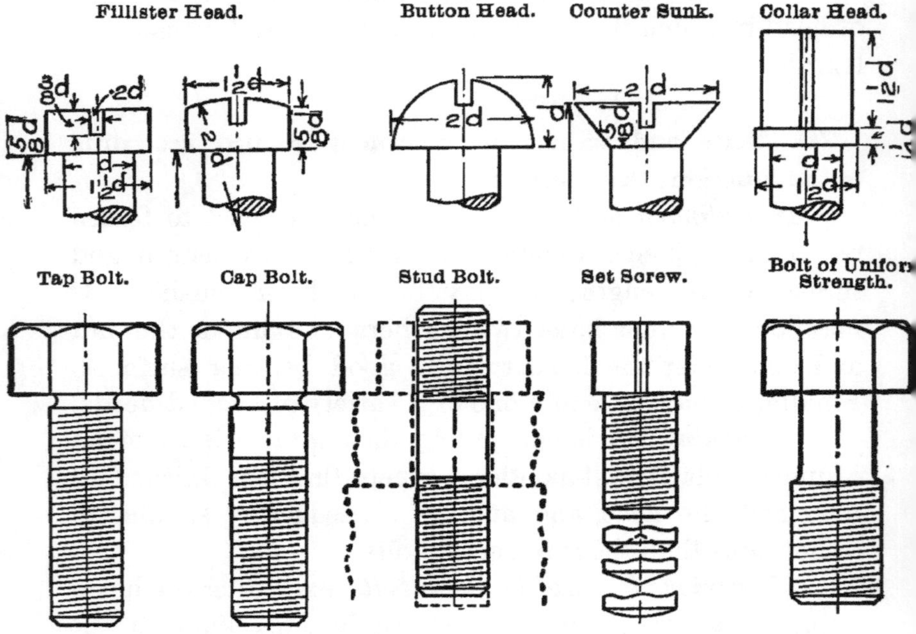

Fillister Head.   Button Head.   Counter Sunk.   Collar Head.

Tap Bolt.   Cap Bolt.   Stud Bolt.   Set Screw.   Bolt of Uniform Strength.

it, a nut is screwed on to hold the whole together. The bolt should not screw down to the bottom of the tapped hole, the depth of the hole should moreover be equal to at least one and one-half times the diameter of the bolt. The smooth part of the bolt should be less in length than the thickness of the piece screwed down over it.

A set screw is one which holds another piece rigid by pressure against its point, such as a screw through a collar holding it fast from turning upon a shaft. There are various forms of points to set screws, generally square heads. They are usually threaded throughout their length.

In nice work on machine bolts, frequently an annular groove is cut underneath the head so as to insure the bolt going in tight and close to the head.

If a bolt is desired of uniform strength, the shank or smooth part will be made equal in diameter to that at the base of the threads.

There are other specially named bolts such as eye bolt, hook bolt, expansion bolt, etc., which can be found described in the hand books. Fig. 49 shows the various forms of bolts above enumerated and which are the more common.

There are other specially named bolts according to the the special uses they are put to, as: *T headed bolts*, having two faces of the head coincident with a square neck of the diameter of the bolt or else tangent to the shank. It is used where there is not room enough to use a square or hexagonal headed bolt. *A hook bolt*, a bolt of square shank and turned at the end to catch on a ledge, used, for example, in fastening hangers to flanged beams. *Expansion bolt*, one used to attach parts to stone, or concrete work; it has various forms, chiefly, tending to spread out at the end and fill a tapered hole larger at the inner end. One form, to go into wood work, is split so that when forced in, it spreads out. *Anchor bolt*, used to hold heavy bodies to stone or concrete foundations, similarly to an expansion bolt, except that it holds by means of a nut and washer, the latter being unusually large in proportion to the diameter of the bolt. *Eye bolt*, having the head end turned into an eye or loop, generally of thickness equal to the diameter of the bolt.

71. A rivet is a permanent fastening usually used to connect two or more thin plates. They are short and the shape of their head gives them their name. The rivet consisting simply of a head and shank, is put in place, either hot or cold, generally the former, and the second head is formed by upsetting with a tool especially constructed. Rivets are now put in place, when in large numbers, by means of a pneumatic riveter which automatically forms the head, when in place. Fig. 50 shows some of the chief forms of rivet heads and the proportions in common use.

FIGURE NO. 50.

Button Head   Snap or Conical Head.   Steeple Pointed Head.   Countersunk Head

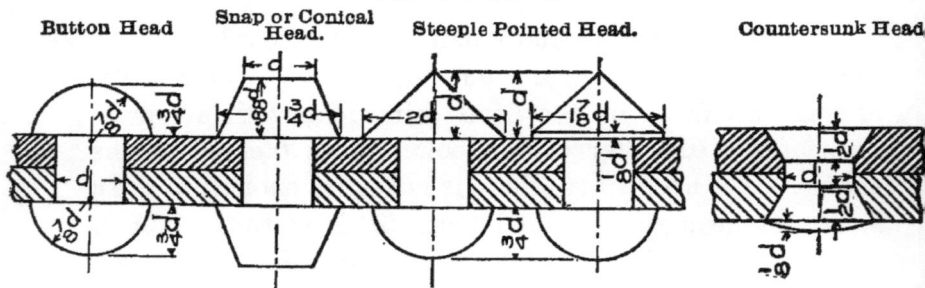

72. General directions for the treatment of problems to follow:

Draw the center lines first and next the centers of arcs and curves tangent to straight lines; that is, make the outline or contour forms, except in the case of very small curves or fillets, determined by the curves to which they are tangent. In construction, the exact location of these centers may be very important. After the centers are located and the curves drawn, enclose the remaining part of the views by straight lines, and so on to the finish.

To be exact in the tangency of curves to straight lines, and of curves to each other, observe that the marks made in the paper by the needle point centers are as small as possible.

Indicate centers for all arcs, other than very small ones like fillets, by at least two center lines crossing each other at right angles at the center, say ¼ of an inch long.

Dotted lines can be made quite explanatory of forms,
The thickness of a dotted line should be no greater than
that of the solid outlines, in fact, it is well to make it
slightly lighter, for the short dots are likely each to be
thicker than a continuous line made with the same setting
of the pen. *A dotted line is not ideally composed of dots,*
but of short, uniform and evenly spaced strokes. The
uniformity of dotting is a thing which sets off a drawing.
Dotted construction should look intelligible always. To
effect this, several points should be observed:

(*a.*) The angles of sharp corners should be connected strokes.

(*b.*) Where the dotted line crosses solid lines,
and is not related to them by identity of form
or material, the dots should not stop at, but
cross, these lines.

(*c.*) Where dotted lines properly end at solid
lines, the dots should touch the lines.

(*d.*) In the dotted lines in either of the above
cases, the dots marking the angles, or those
crossing solid lines or meeting to end at solid
lines, can well be made longer than the dots
of the rest of the line, even twice the length.

Dotting, in brief, should clearly define the forms by
accentuating with larger dots the salient things to be
brought out.

73. Miscellaneous problems.

Problem 1. Draw a plan and elevation of the form
shown in Fig. 51, adding the dimensions as given and
placed in similar positions on the views.

FIGURE NO. 51.

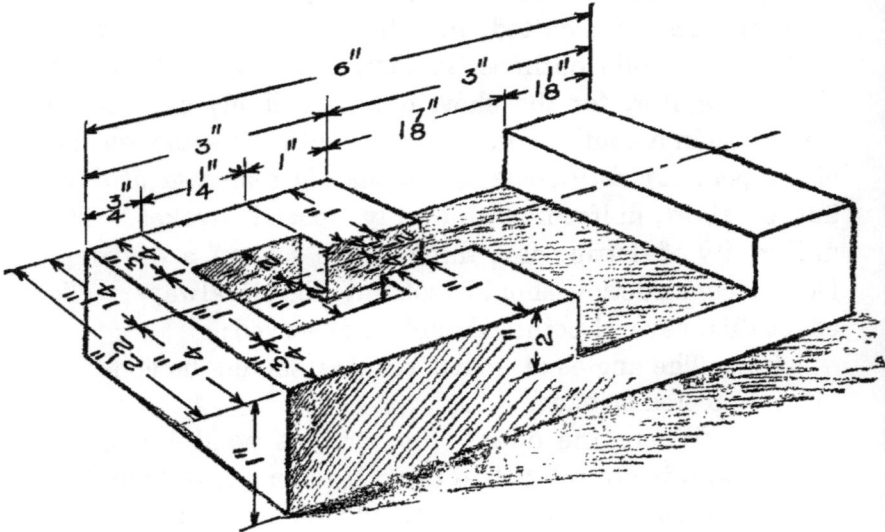

Draw first the plan view, laying out the center line, scaling the length over all, and the width of the view.· Next, lay off the shorter horizontal distances, preferably upon the center line, marking continuously from the scale, then the vertical measurements, either side of the center line at the points marking the horizontal distances. Line in the view, horizontal lines first and then vertical lines.

Lay out the elevation in a similar manner to the plan. Put on the limiting lines, dimension lines, arrow heads and figures in the order named. The dimensions occupy relatively the proper place upon the figures that they should in the working drawing.

In inking, rule the horizontal lines of all the views first, beginning at the top and going down. Rule the vertical lines next, including dotted lines and invisible edges. Rule the limiting and dimension lines next, and lastly put in arrow heads and dimension figures in the order named.

Problem 2. Draw a plan and elevation of the form shown in Fig. 52, adding the dimensions as given.

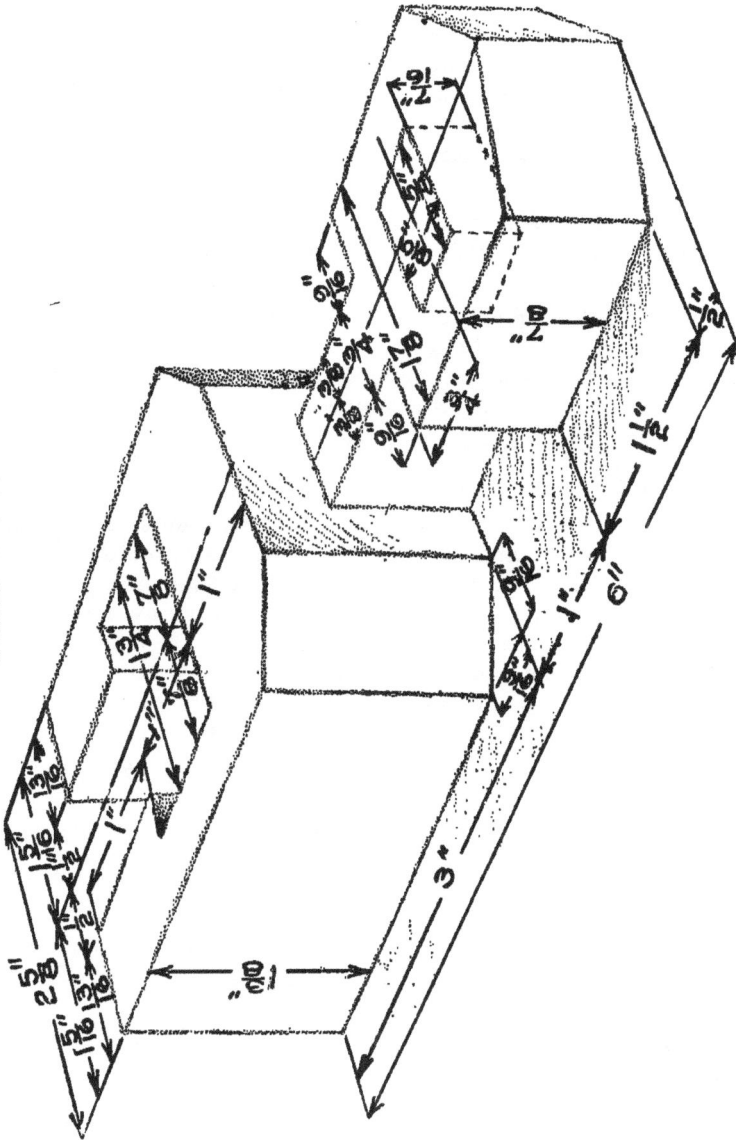

FIGURE No. 52.

Draw first the plan, laying out the center line, scaling the length over all, and the width of the view. Next, lay off the shorter horizontal distances, preferably upon the center line, marking continuously from the scale, then the vertical measurements, either side of the center line at the point marking the horizontal distances. Line in the view, horizontal lines first and then vertical lines.

Lay out the elevation in a similar manner to the plan. Put on limiting lines, arrow heads and figures in the order named.

In inking, rule the horizontal outlines of all the views first, beginning at the top and going down. Rule the vertical outlines next, including dotted lines or invisible edges. Rule the limiting and dimension lines next, and lastly, put in arrow heads and dimension figures in the order named.

Problem 3. Draw plan, elevation and end view of the form shown in Fig. 53, adding the dimensions given.

Draw first the plan of the form, laying out center lines, scaling the length over all and the width of each end, then enclosing by lines. Draw the elevation next, and then either end. After this put ont he dimension lines, limiting lines and arrow heads, lastly the dimension figures.

In inking, rule the horizontal lines of all the views first, beginning at the top and going down. Rule the vertical outlines next, finally the oblique lines. Next rule the limiting and dimension line, and lastly put in arrow heads and dimension figures in the order named.

Problem 4. A flange for outlet and inlet pipes for a hydraulic riveting machine. See Fig. 54. The pipes screw into the tapped holes shown. Copy the views shown, making full size. This figure shows the proper weight of line for a working drawing.

Draw the main center lines of the upper or plan view first, continuing the vertical one down to the elevation. Next draw the minor center lines of the plan, then close in the plan, smaller circles first, then the larger and the straight lines last. Draw the elevation from the plan, making the upper and lower horizontal

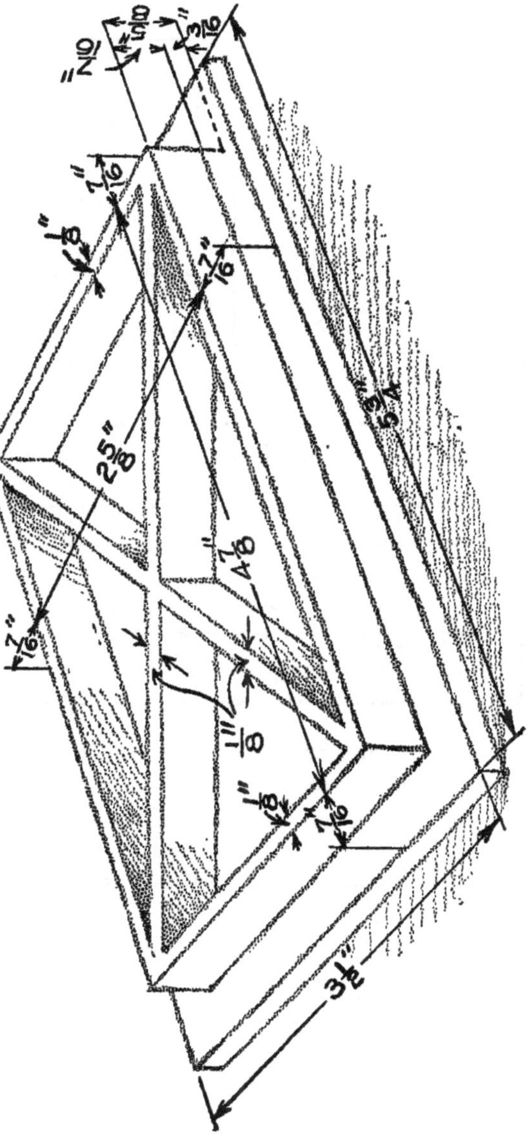

FIGURE NO. 58.

FIGURE NO. 54.

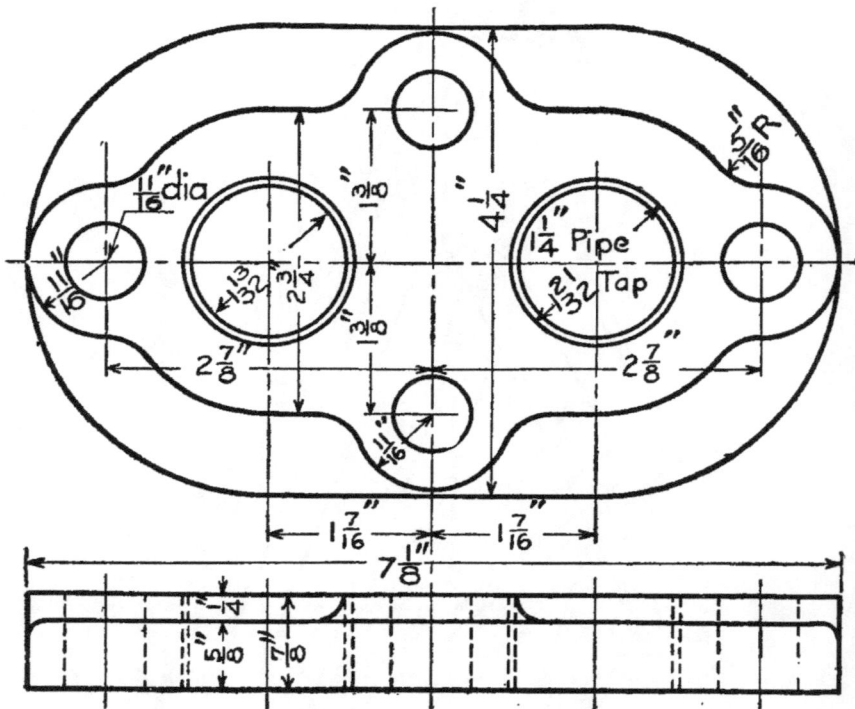

limiting lines first, then projecting down from the plan. **Put the dimensions on as on the figure.**

When ready to ink, put in the small arcs with the bow pen, then the larger with the compass, and lastly the straight lines. Put in the limiting lines, dimension lines, arrow heads, and figures in the order named.

Problem 5. Draw an under plan, an elevation and an end view of the form shown in Fig. 55, adding the dimensions as given.

Draw first the plan view, laying out the center lines, vertical and horizontal, scaling the length over all and the width of the view. Next locate, centers for and draw all arcs of circles.

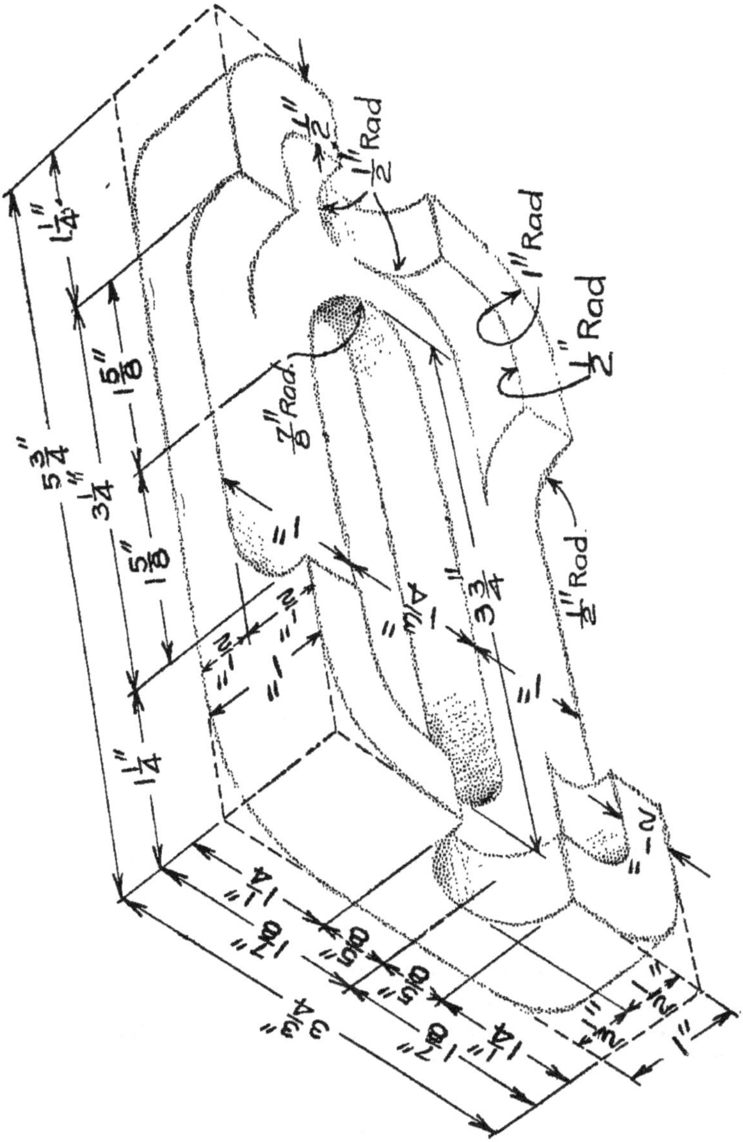

FIGURE NO. 55.

connecting same by tangent and enclosing lines. Draw the
elevation, then the end view from the plan following same general
procedure. Put in the dimensions as on the figure in the same
relative positions, except show radii upon a radial dimension line,
lettered R or Rad.

When inking put in the arcs of circles first, in all views, then
the straight lines, horizontal and vertical in turn. The limiting
lines, dimension lines, arrow heads and figures go in last in the
order named.

Problem 6. Draw a plan, an elevation and an end
view of the form shown in Fig. 56, adding the dimensions
as given.

Draw first the end elevation, which is a view looking toward
the left, laying out the center lines, scaling the length and height
over all and constructing either side of the center line. Locate
the centers for, and draw all arcs the first lines drawn, connecting
same by tangent and enclosing lines. Draw the front elevation
and then the plan next following the same general procedure
except that the plan will have two axes of symmetry. Put in
the dimensions as on the figure in the same relative positions,
except show the radii upon radial dimension lines, lettered R or
Rad.

When inking put in the arcs of circles first in all views, then
the straight lines, horizontal and vertical in turn. The limiting
lines, dimension lines, arrow heads and figures go in last in the
order named.

Problem 7. A cross section of a P. R. R. standard
100-lb. rail, see Figure 57. Draw the view from the
sketch, full size, to the dimensions shown. It is an
exercise in the tangency of curves.

Draw the vertical center line first, then the centers for the
arcs designating the web, then the crown of the rail, and the base
and so on, locating centers in the penciling carefully by free-hand
circles around the intersections of the construction lines. Put on
the dimensions as shown.

When ready to ink, put in the small arcs with the bow pen,
then the larger with the compass and lastly, the straight lines.

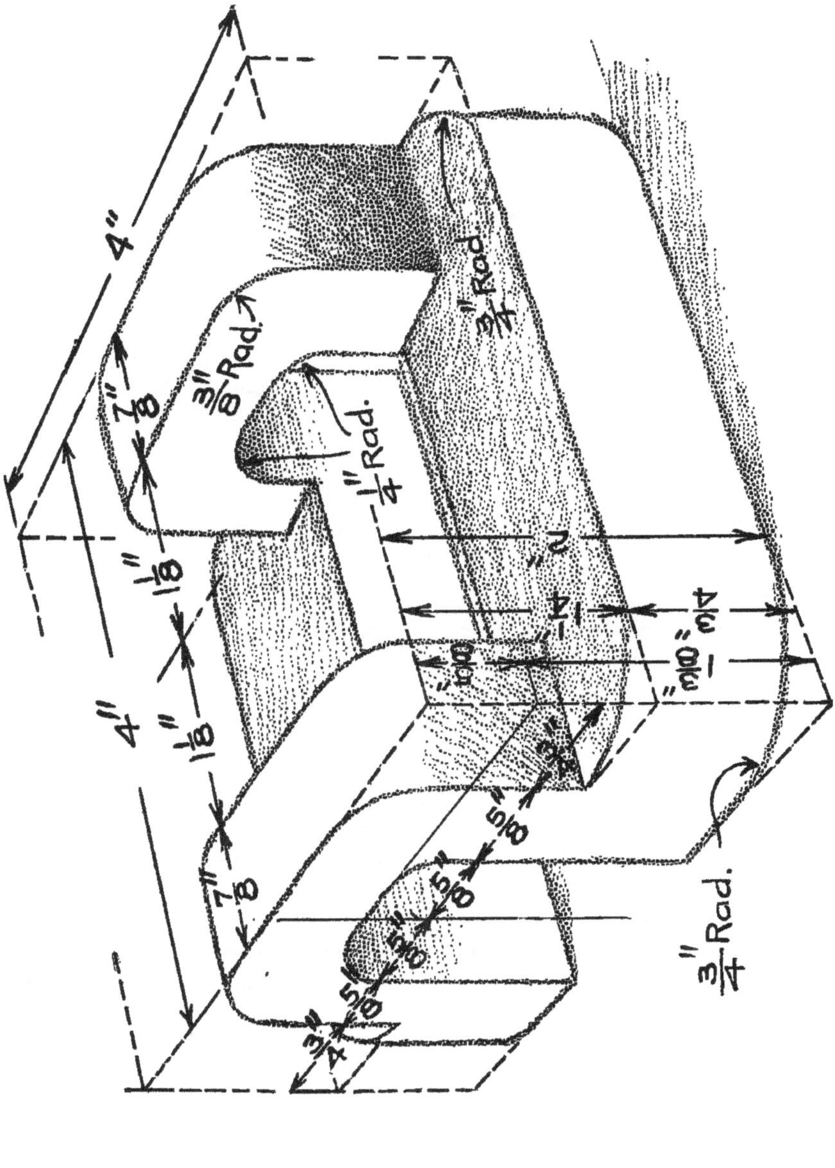

FIGURE No. 56.

FIGURE No. 57.

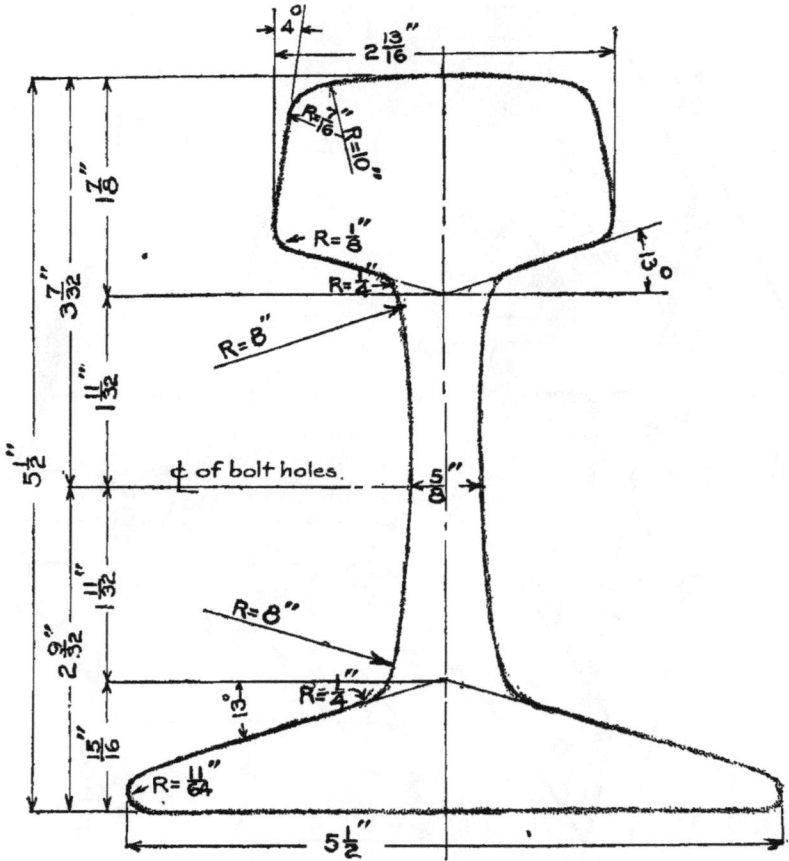

Show the plane of the section by regularly spaced lines, at 45° to
the horizontal and about one-sixteenth of an inch apart at the
least, the weight of the line to be slightly lighter than the outline.
Put in the limiting lines, dimension lines, arrow heads and figures
in the order named.

Problem 8. Make the necessary working drawing
views to show the stop lever in Fig. 58.

Draw the horizontal and oblique center lines of both views
first. Next, draw the central boss and left hand end of the lower
view or elevation. Then draw the fork of the plan view and
project down to the elevation. Finish the elevation then the plan.
Put the dimensions on to correspond with their general positions
on the sketch.

When ready to ink, draw the circular arcs first, then the horizontal, vertical and oblique lines in order. Put in limiting lines,
dimension lines, and arrow heads and figures in the order named.

Problem 9. An overhung crank for an engine. Draw
a front and side view from the sketch in Fig. 59, full size.

FIGURE No. 59.

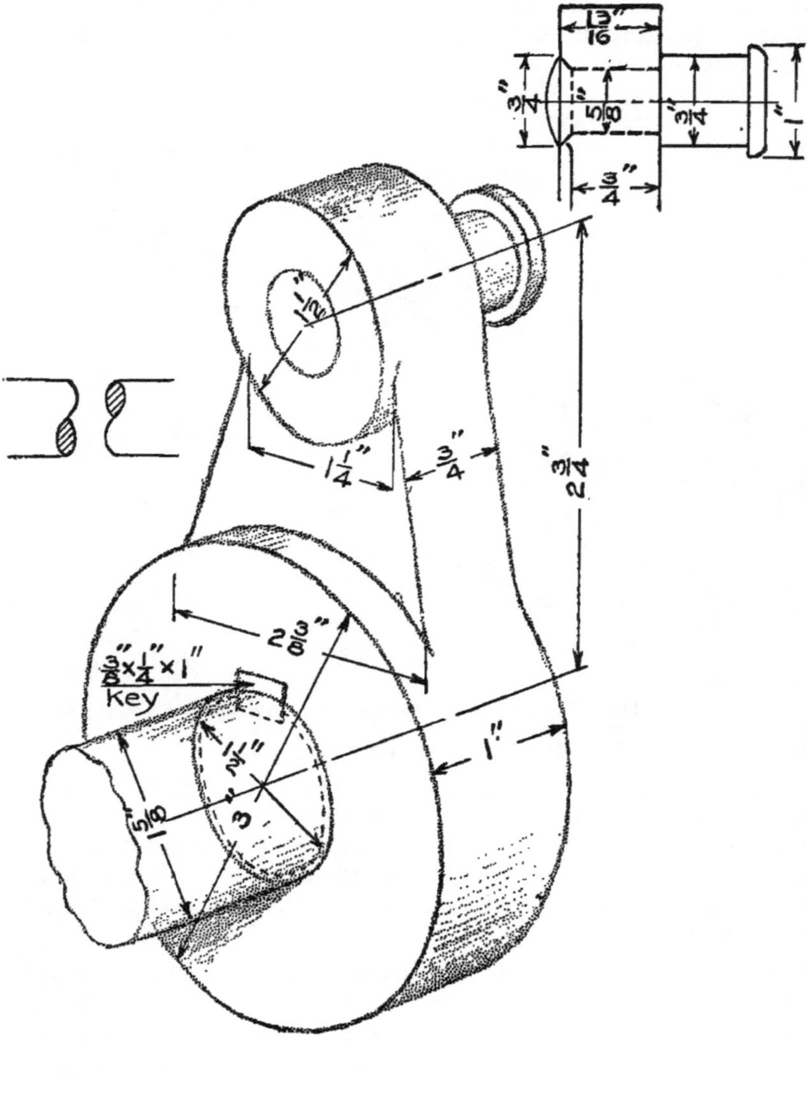

Call the front view the one looking in the direction of the shaft; this should be drawn first.

Note the conventional method of showing a break in a shaft by aid of the small detail at the left of the figure. The curved part is sometimes made with the curved rule, but this is not at all necessary. The small detail at the right shows the demensions which could not be represented on the perspective sketch.

The sectioning, or conventional method of showing the cut surface should, in all cases, be ruled by equally spaced lines somewhat lighter than the general outlines, about one-sixteenth of an inch apart in this problem.

Problem 10. Draw the front view and side view of the rocker arm as shown by perspective sketch in Fig. 60. Call the front view the one looking in the direction of the shaft at the top.

This form will show the advantage of developing parts of several views simultaneously. The front view, including the split collar, and what is below, is best drawn first; that which is above the split collar, including the tightening bolt, should be drawn in the end view first. Put in small rounded edges or fillets last in pencil, but first in inking.

Problem 11. A discharge cap for a pump, see Fig. 61. Make the views shown by sketch to a scale of three inches to the foot, and in proper position, render the right hand half of both side and end views in section.

*Problem 12. A brake shoe, see Fig. 62. Make a plan, elevation and side view of the form shown by sketch, to a scale of three inches to the foot. Also make a section

---

* From Rautenstrauch & Williams' 'Machine Drafting.'

FIGURE No. 60.

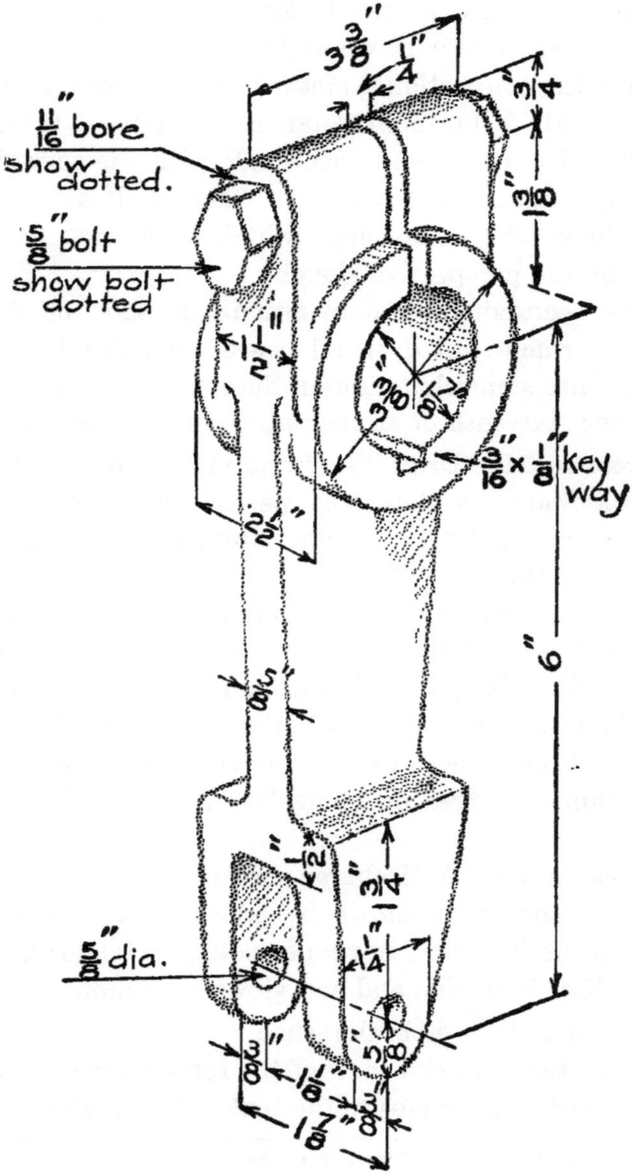

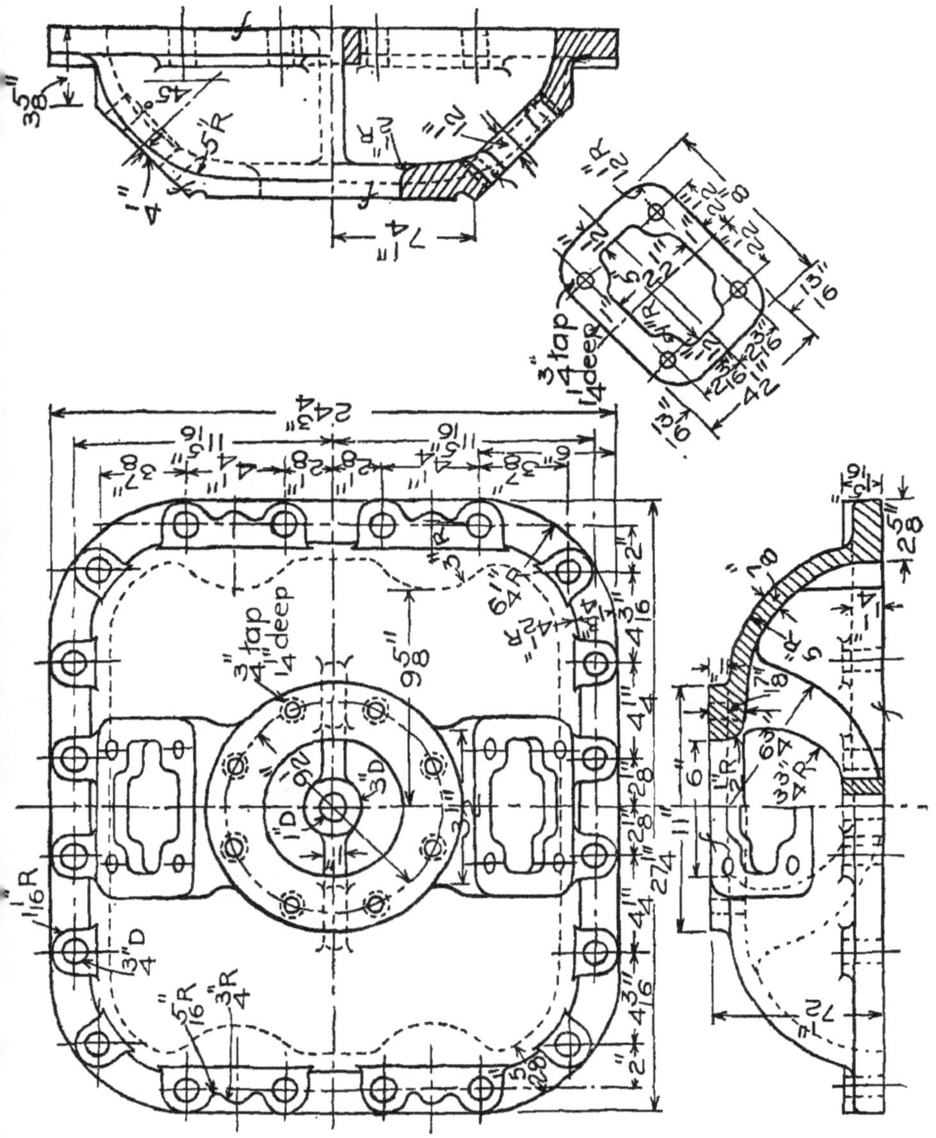

FIGURE No. 62.

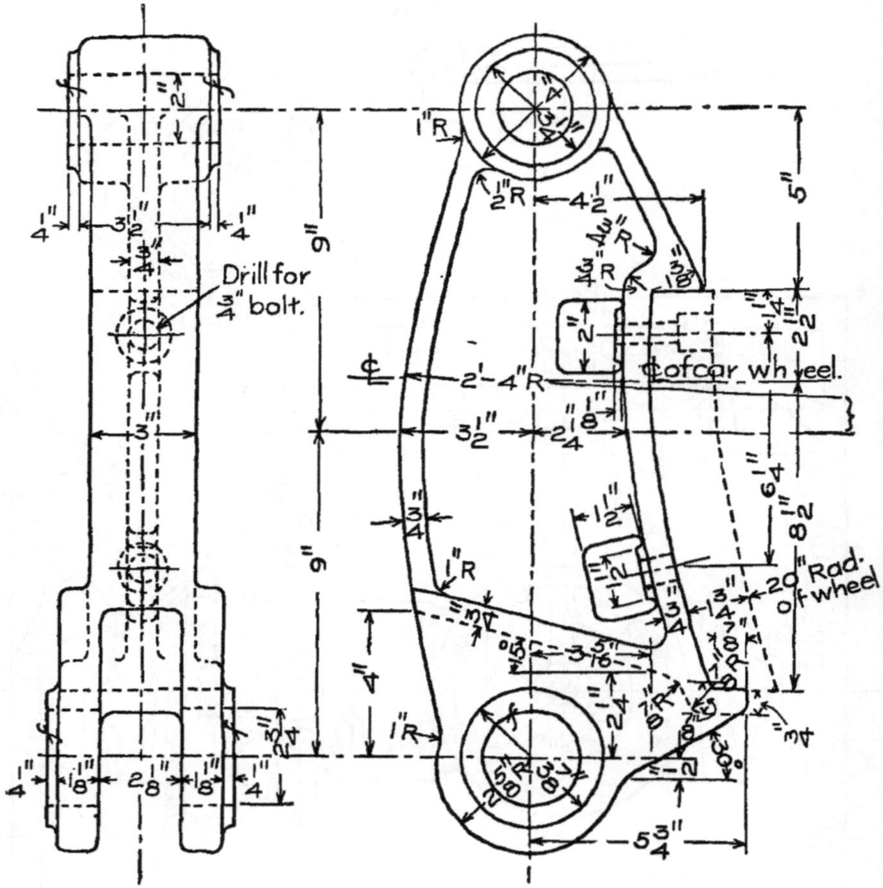

FIGURE NO. 68.

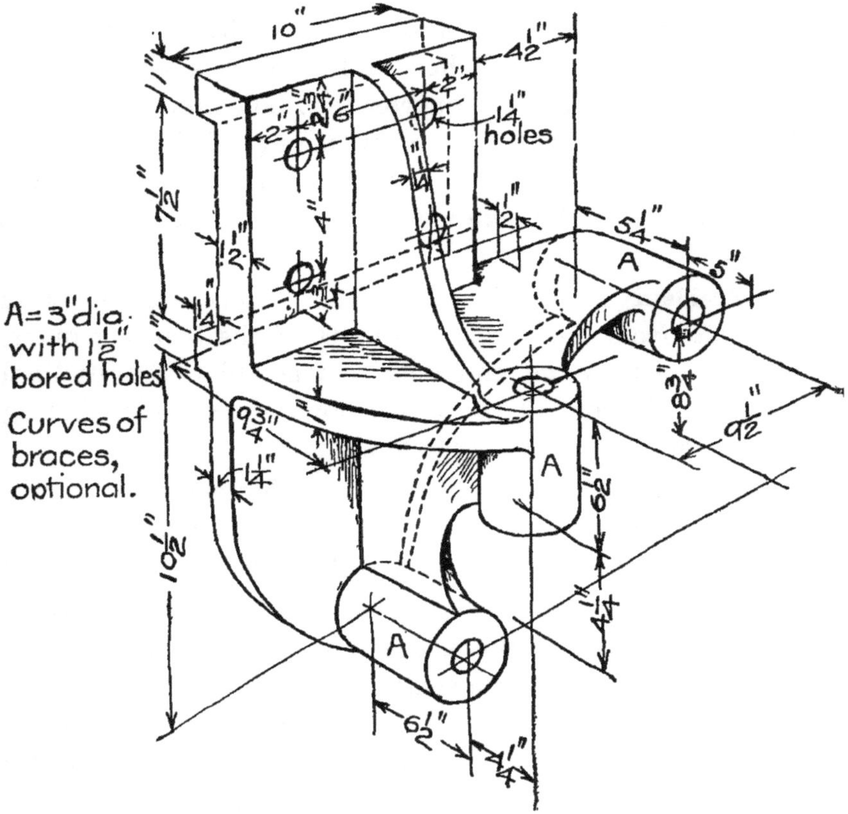

looking downward at the center line of car wheel. Put on all dimensions shown.

Problem 13. Make a plan, elevation and end view of the bracket shown in Fig. 63, to a scale of three inches to the foot. Put on all necessary dimensions.

Problem 14. Make a working drawing of the pillow block as sketched in Fig. 64, to a scale of four inches to the foot. Make also a longitudinal section through the center. Put on all dimensions as shown.

Problem 15. Make a drawing of the cylinder head and stuffing box as sketched in Fig. 65 to a scale of half size. Draw in the three-quarter inch bolts and nuts on the left hand half of the front view, facing all nuts the same way. Draw the front view first, taking sizes where necessary from the side view. Put on all the dimensions as shown.

Problem 16. A bench grinder. Make an elevation complete from the sketch in Fig. 66 of the bench grinder, also a plan view and end view and complete longitudinal section. Put on all the dimensions shown, distributing them properly between the views.

Problem 17. A bench grinder. Make a set of working drawings complete of all parts of the bench grinder as sketched in Fig. 66, and to a suitable scale. Put on all the necessary dimensions and tabulate a bill of materials.

Problem 18. Make a drawing of the engine bed as sketched in Fig. 67, to a scale of one-fourth size. Put on all dimensions, where possible in their proper place, but if there is not room, place to one side and refer to them by direction lines as shown here and there in the sketch.

Make a central longitudinal section of the bed.

Problem 19. Make a drawing of the 'Standard Pile Bridge' construction shown by sketch in Fig, 68a and 68b, and as used by the C. B. & Q. R. R. Make it to a scale of one-fourth inch to the foot, and arrange the views appropriately.

FIGURE NO. 64.

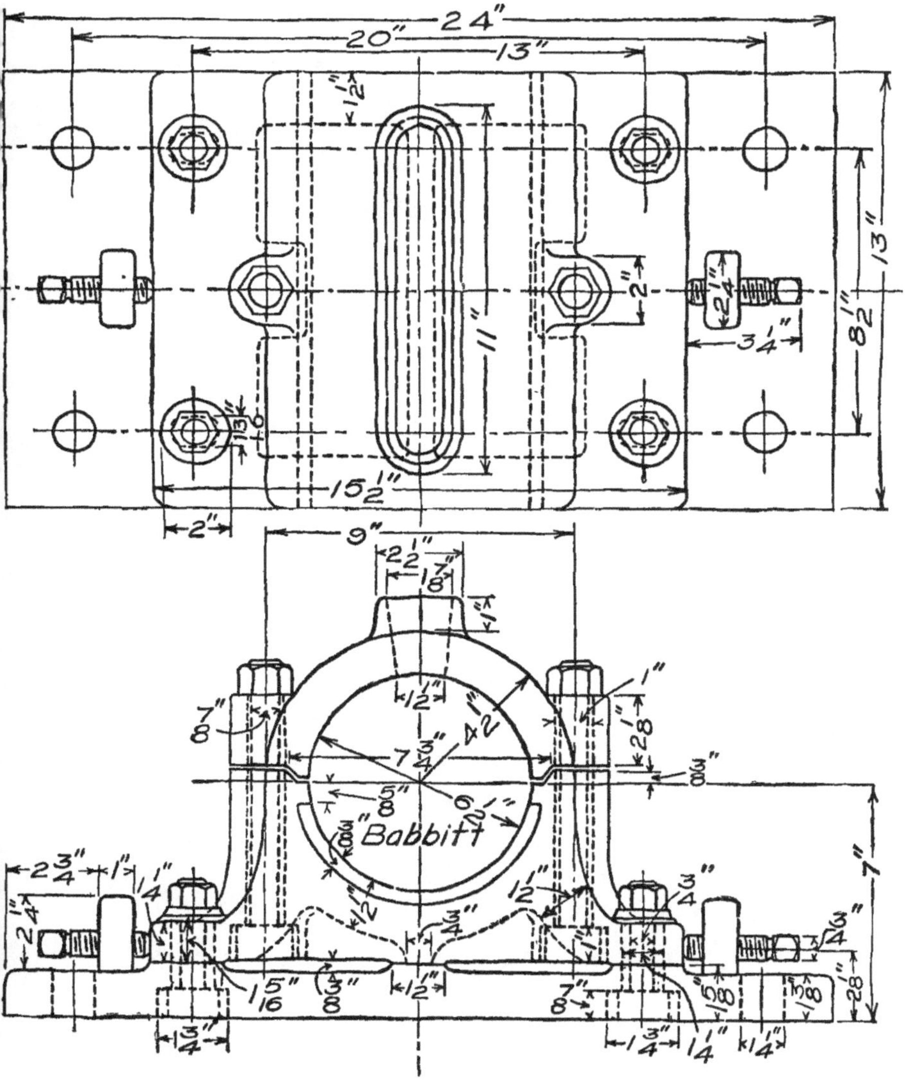

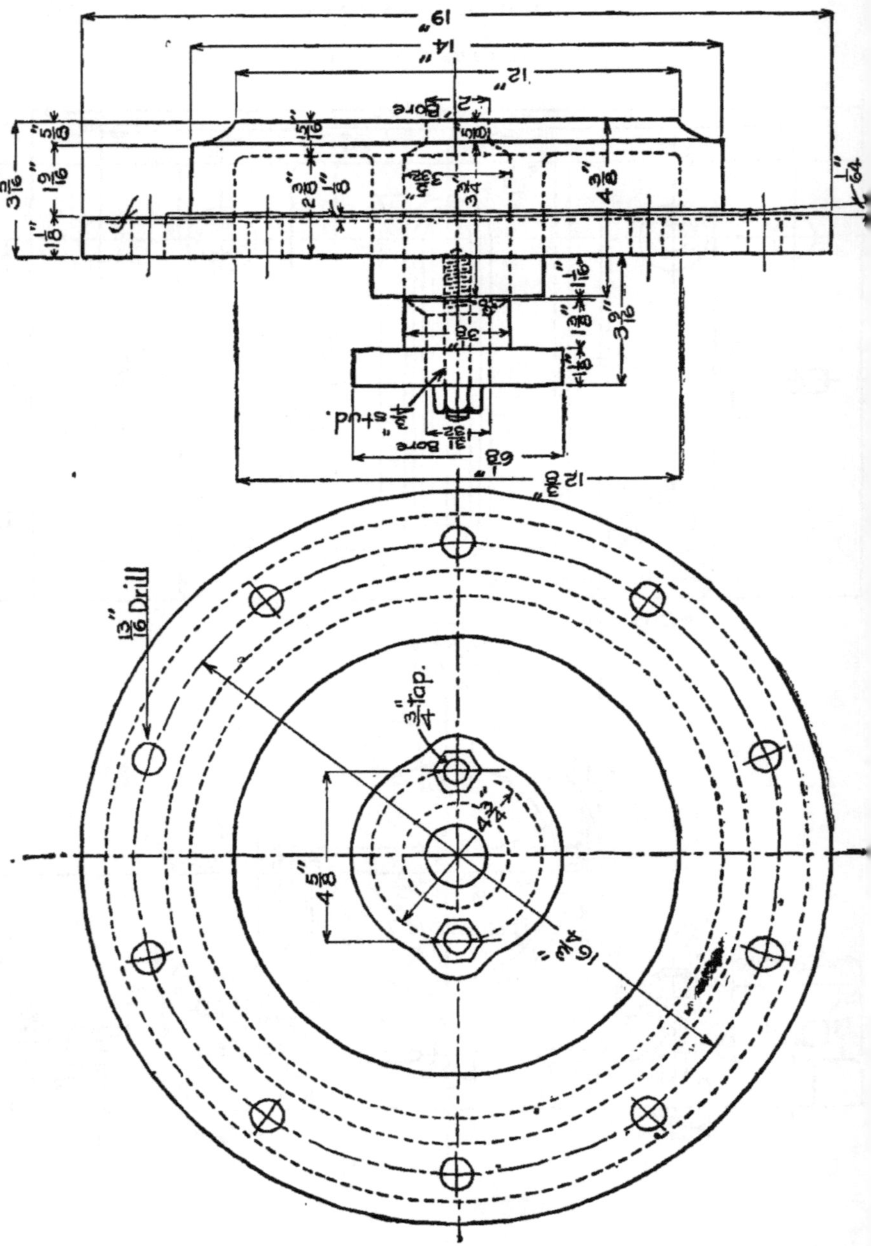

FIGURE NO. 65.

FIGURE No. 66.

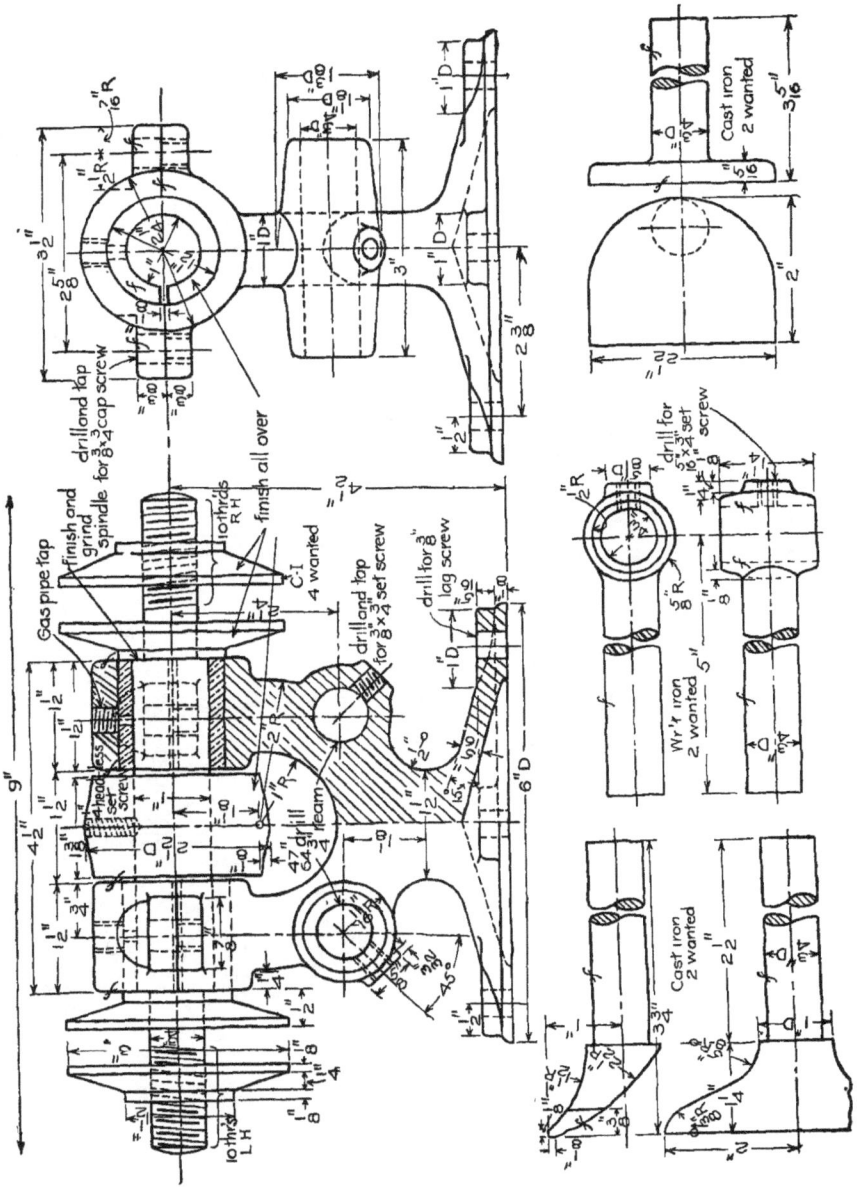

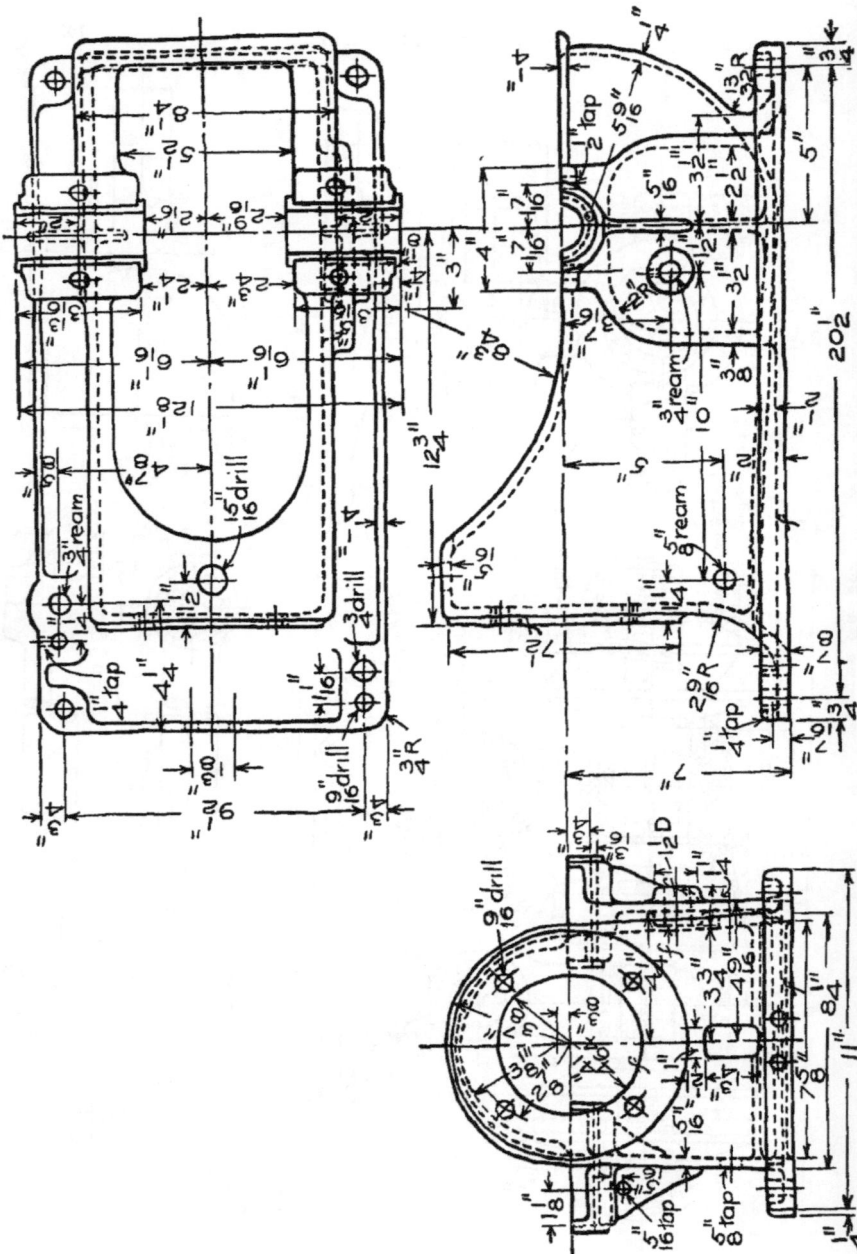

FIGURE No. 67.

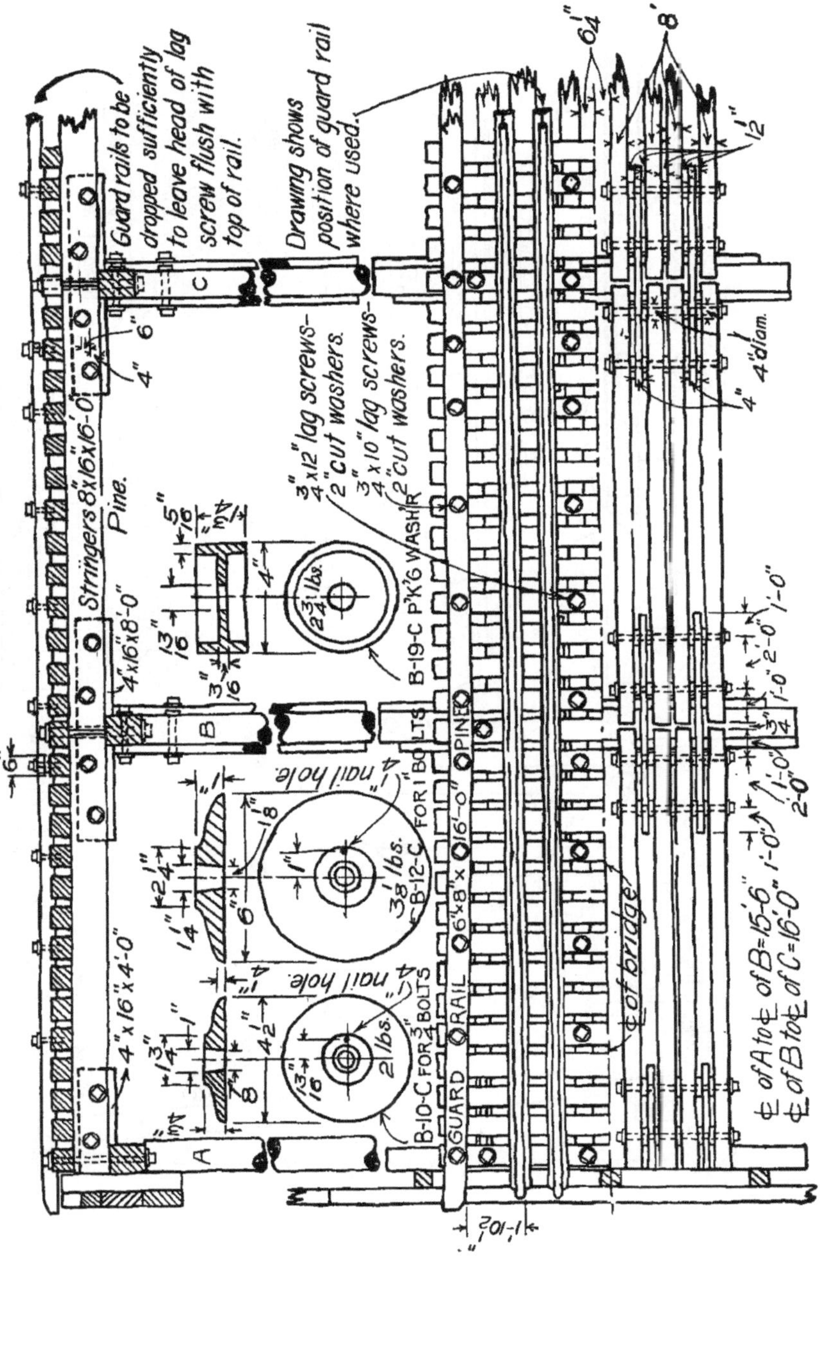

FIGURE No. 68a.

Guard rails to be dropped sufficiently to leave head of lag screw flush with top of rail.

Drawing shows position of guard rail where used.

Stringers 8"x16x6'-0" Pine.

4"x16'x8'-0"

4"x16"x4'-0"

3"x12" lag screws—2 cut washers.

3"x10" lag screws—2 cut washers.

B-19-C P'K'G WASH'R

B-12-C

B-10-C FOR 3" BOLTS

FOR 1" BOLTS

GUARD RAIL

6"x8"x16'-0" PINE

4 nail hole

4 nail hole

2⅜ lbs.

3⅜ lbs.

2 lbs.

C of bridge

C of A to C of B=15'-6"

C of B to C of C=16'-0"

FIGURE No. 68b.

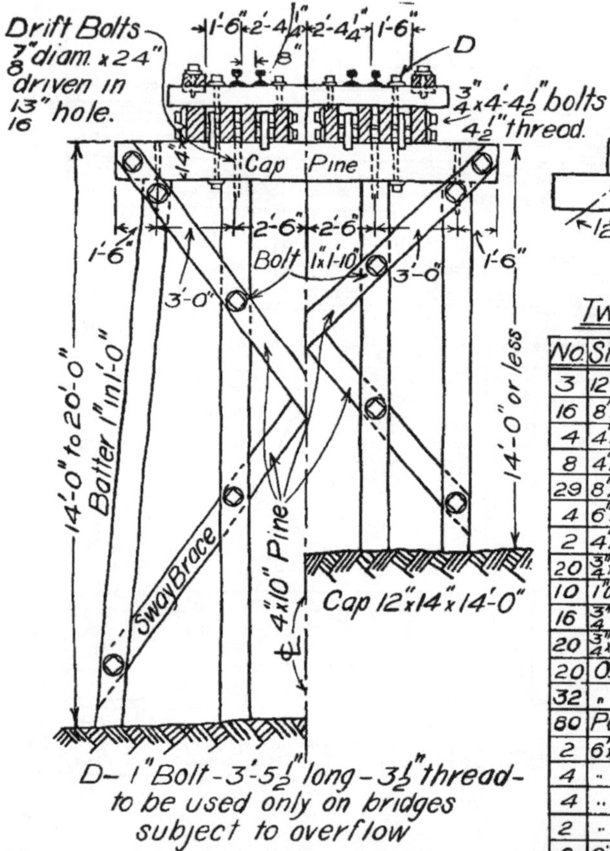
Drift Bolts $\frac{7}{8}$" diam. x 24" driven in $\frac{13}{16}$" hole.
1'-6" 2'-4$\frac{1}{4}$" 2'-4$\frac{1}{4}$" 1'-6"
8"
D
$\frac{3}{4}$"x4'-4$\frac{1}{2}$" bolts 4$\frac{1}{2}$" thread.
Cap Pine
1'-6"
2'-6" 2'-6"
Bolt 1"x1'-10"
3'-0"
1'-6"
14'-0" to 20'-0"
Batter 1" in 1'-0"
14'-0" or less
Sway Brace
4"x10" Pine
Cap 12"x14"x14'-0"
D– 1" Bolt-3'-5$\frac{1}{2}$" long-3$\frac{1}{2}$" thread-
to be used only on bridges
subject to overflow

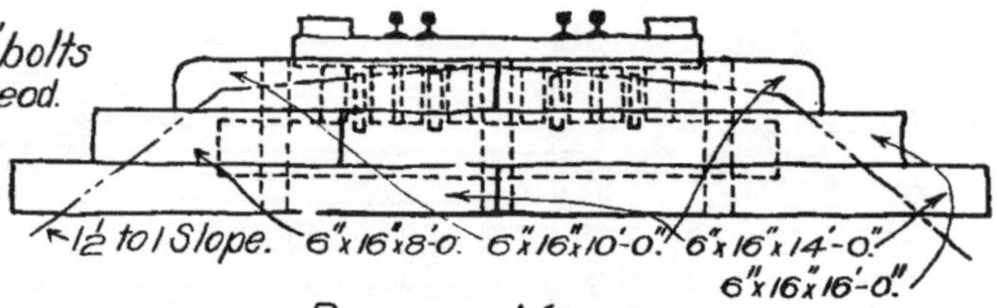
1$\frac{1}{2}$ to 1 Slope.
6"x16"x8'-0.
6"x16"x10'-0."
6"x16"x14'-0."
6"x16"x16'-0."

BILLS OF MATERIAL

TWO END PANELS.

No.	Size	Length	
3	12"x14"	14'-0"	Caps
16	8"x16"	16'-0"	Stringers
4	4"x16"	8'-0"	
8	4"x16"	4'-0"	
29	8"x8"	10'-0"	Ties
4	6"x8"	16'-0"	Guard rail
2	4"x10"		Sway braces
20	$\frac{3}{4}$"x10"	Lagscrew-2" cut washer	
10	1" diam.	1'-10" U H.	Bolts-2" th'd
16	$\frac{3}{4}$" "	4'-4$\frac{1}{2}$" "	Bolts " "
20	$\frac{3}{4}$"x12"	Lagscrew-2" cut washer	
20	O.G. washers-B-12-C		
32	" "	"	-B-10-C
80	Packing washers-B-19-C		
2	6"x16"	16'-0"	Old Stringers
4	" "	14'-0"	" "
4	" "	10'-0"	" "
2	" "	8'-0"	" "
6	6"x8"		

INTERMEDIATE PANEL.

No.	Size	Length	
1	12"x14"	14'-0"	Cap
8	8"x16"	16'-0"	Stringers
4	4"x16"	8'-0"	
14	8"x8"	10'-0"	Ties
2	6"x8"	16'-0"	Guard rail
2	4"x10"		Sway braces
10	$\frac{3}{4}$"x10"	Lagscrew-2" cut wash	
10	1" diam.	1'-10" U H.	Bolts-2" th'd
8	$\frac{3}{4}$" "	4'-4$\frac{1}{2}$" "	Bolts- " "
10	$\frac{3}{4}$"x12"	Lagscrew-2" cut wash	
20	O.G. washers-B-12-C		
16	" "		B-10-C
40	Packing washers-B-19-C		
4	$\frac{7}{8}$" diam.	24"	Drift Bolts

Problem 20.   Make a drawing of the 'Fixed End Casting' used by the Illinois Central R. R. on their standard 177'—6'' 'Through Pin Span' and as sketched in Fig. 69. Make it to a scale of one-eighth size.   Put on all dimensions shown.

Problem 21.   Locomotive piston head.   Make a working drawing from the sketch in Fig. 70 to a scale of six inches to the foot.

Problem 22.   Make a drawing of the fly wheel shown by sketch in Fig. 71, to a scale of one and one half inches to the foot.   Draw at least half the wheel in both views.

*__Problem 23.__   Make a drawing of the locomotive driving wheel as sketched in Fig. 72, to a scale of two inches to the foot.   Make also a section of the wheel at right angles to that shown.

Problem 24.   Bench vice.   Make a complete set of working drawings from the sketches in Fig. 73, also a set of assembly views.   The figure illustrates arrangement in working drawing.

---

* From The Locomotive Dictionary.

FIGURE No. 69.

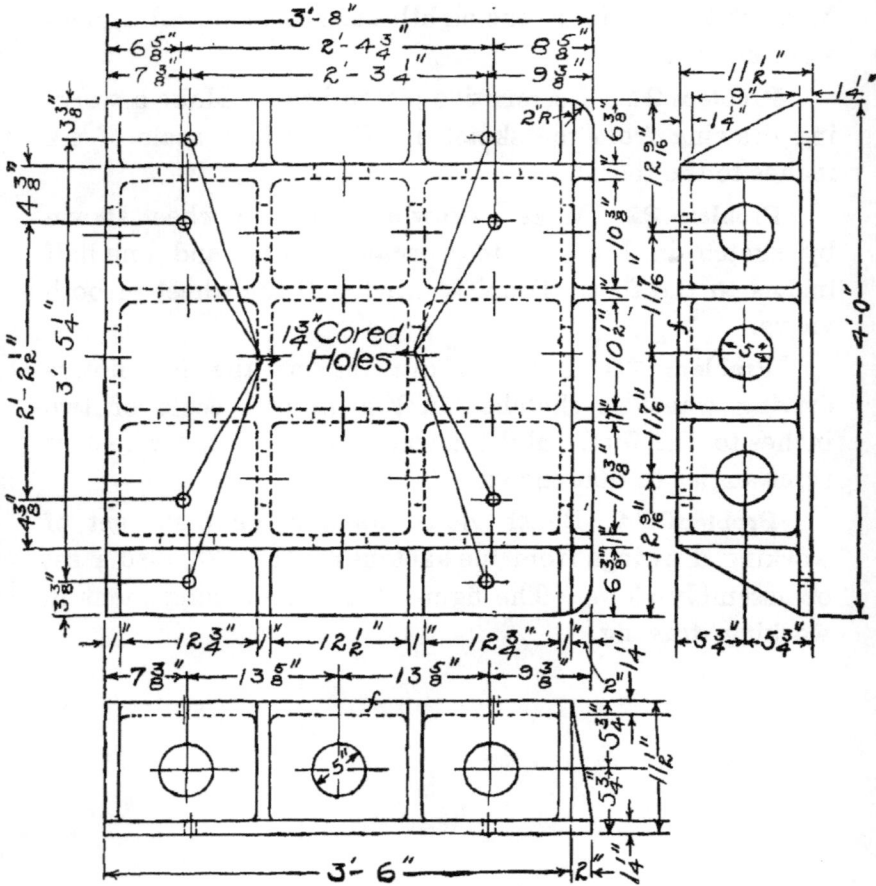

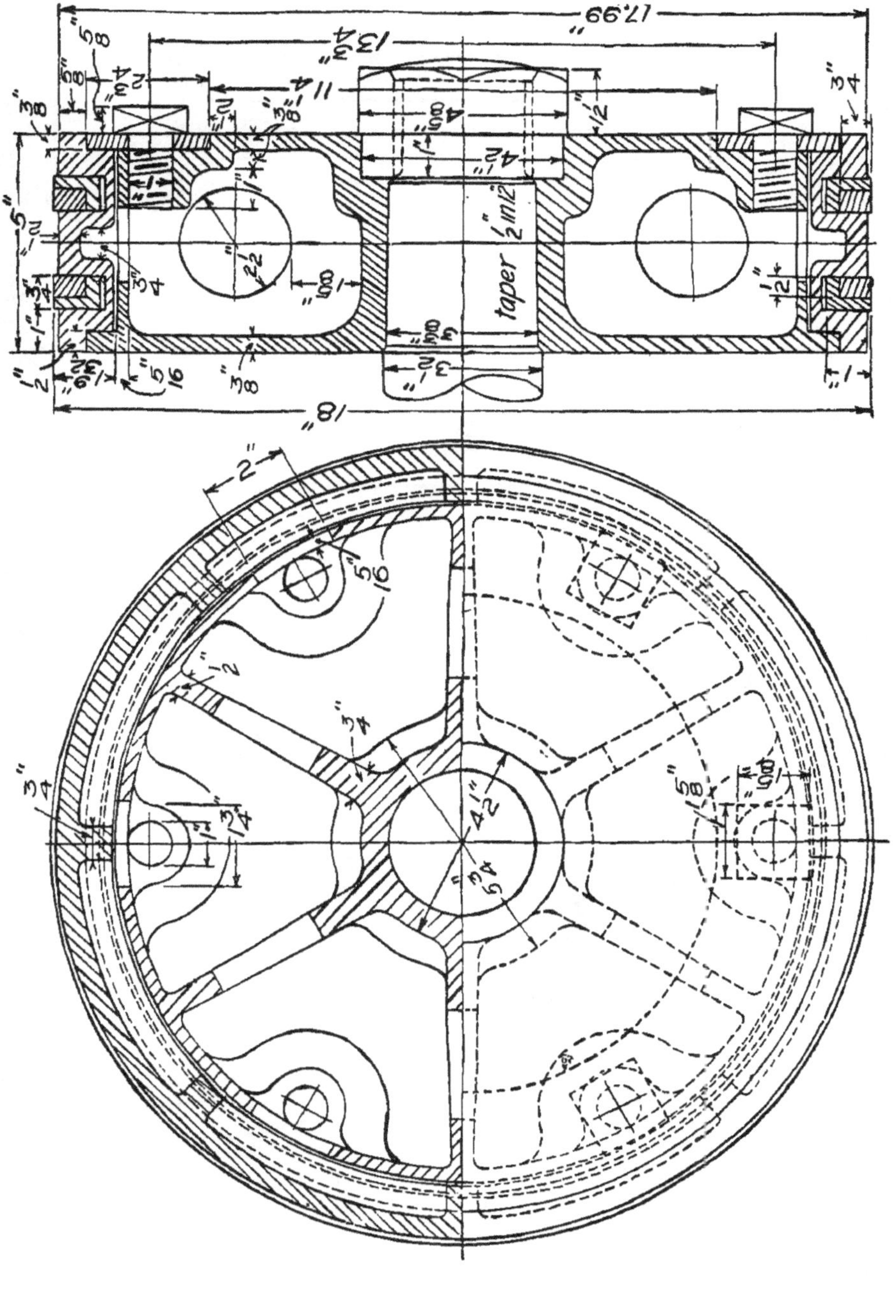
17.99"
18"
taper
2"

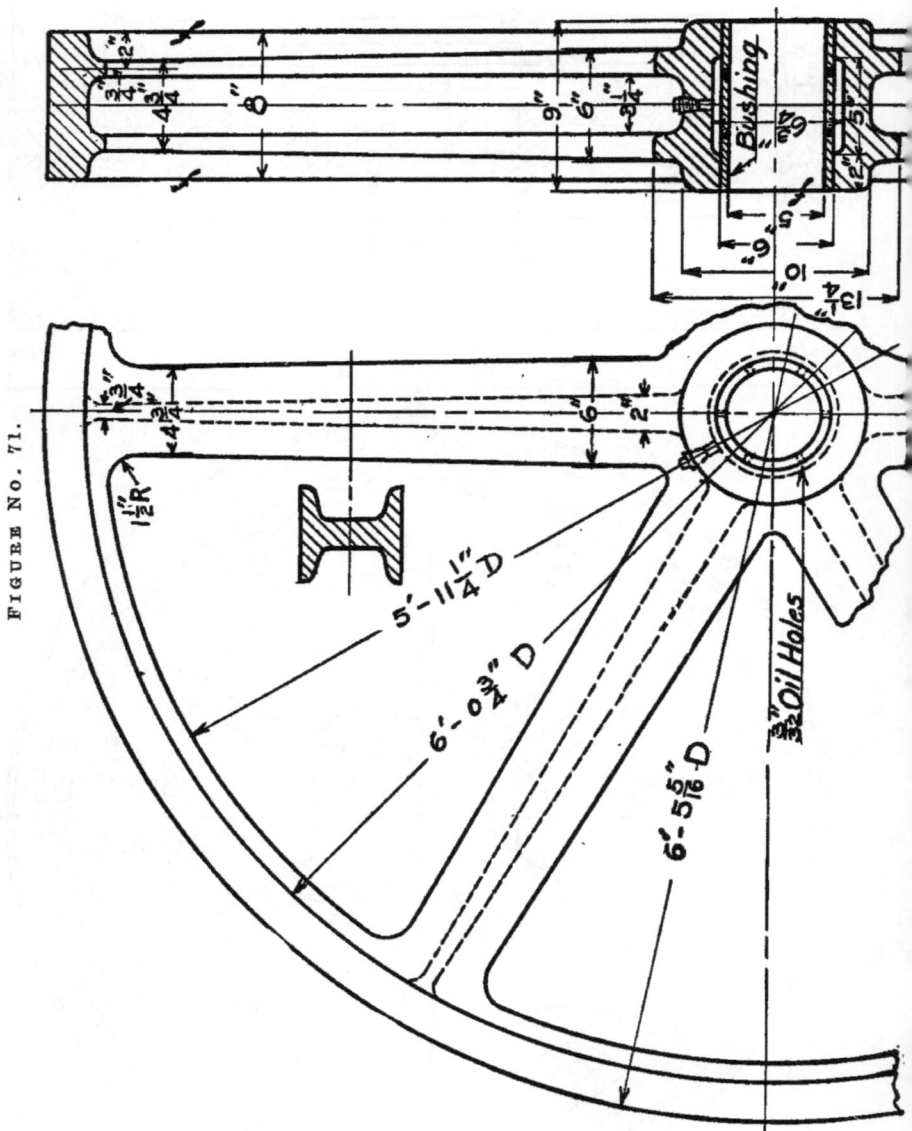

FIGURE No. 71.

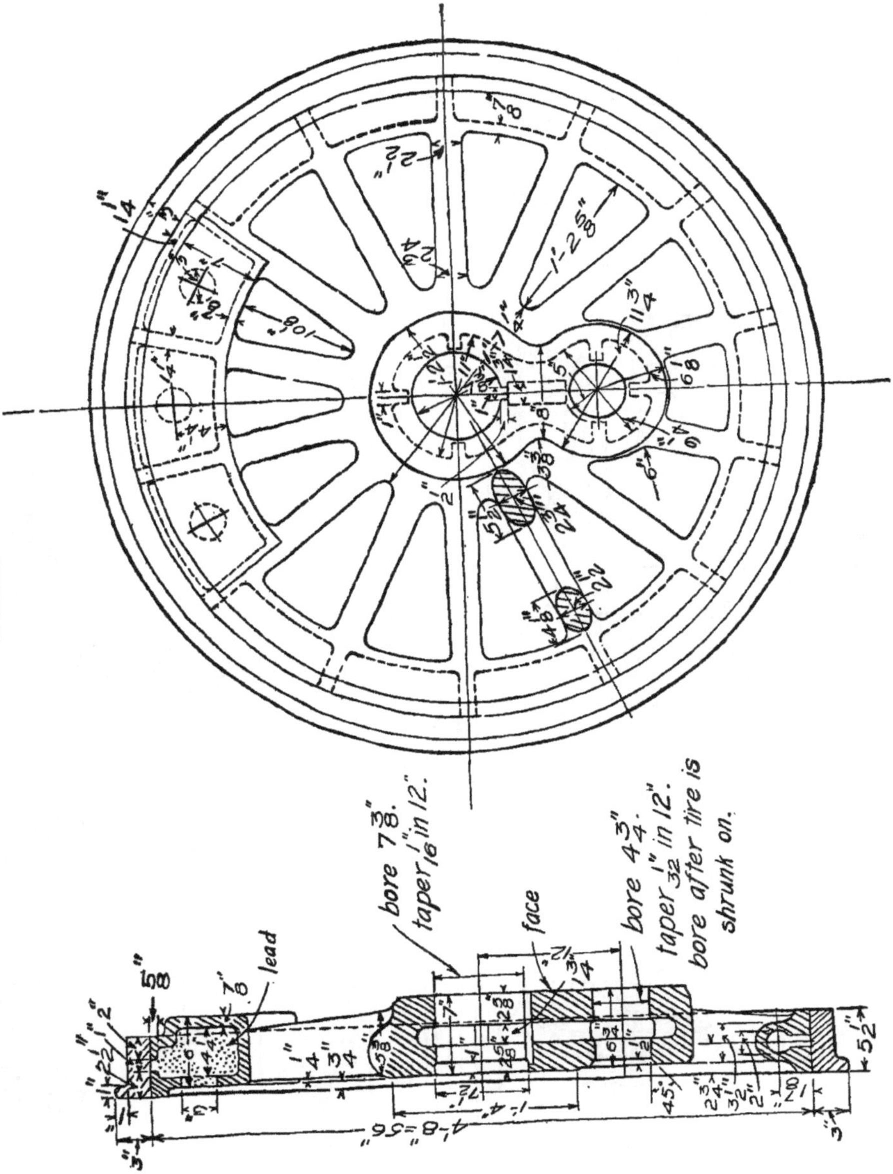

FIGURE NO. 72.

bore $7\frac{3}{8}$". taper $\frac{1}{16}$" in 12".

face

bore $4\frac{3}{4}$". taper $\frac{1}{32}$" in 12". bore after tire is shrunk on.

lead

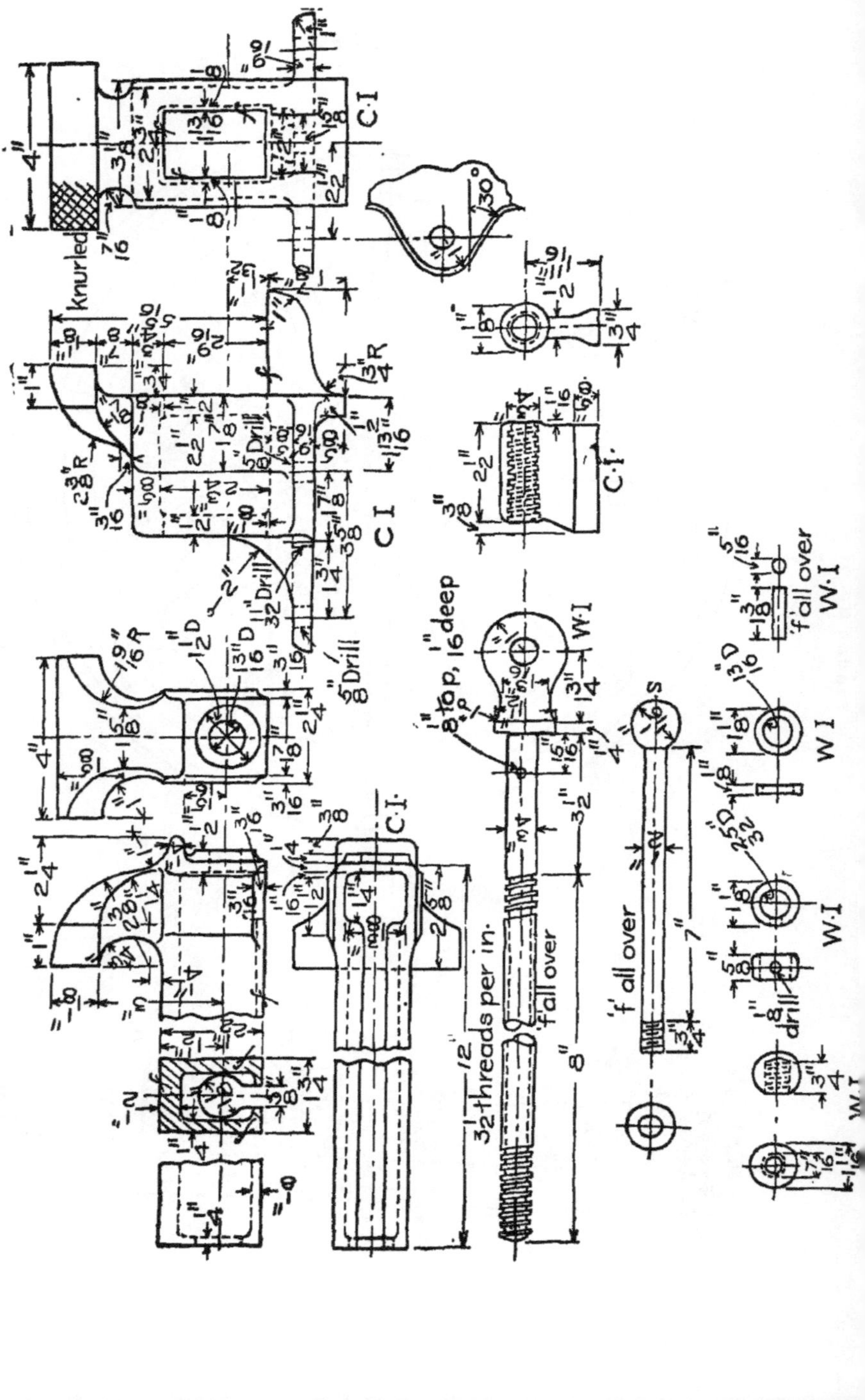
Knurled
C.I
C.I
C.I.
W.I
W.I
W.I
f all over
f all over
f all over
32 threads per in.
tap, deep
Drill
Drill

FIGURE No. 78.

CHAPTER VI.

GEOMETRICAL DRAWING.

74. Geometrical Drawing vs. Mechanical Drawing.

Use all the mechanical aids that are available to get constructive results in a drawing, provided the processes do not involve too great an expenditure of time. Geometrical processes, by their very multiplicity of steps, open the way for errors, and are practically more likely to result in faulty construction than if mechanical aids were used. But there are a few fundamental processes which every draftsman ought to known and there are some mechanical equivalents, also. As for the many geometrical constructions of a more or less simple form, the student is recommended to consult the various hand books.

Parallel lines are more easily obtained by mechanical methods. The T sq. and also the T sq. and triangles taken together, show the simplest cases. For others use either a straight edge and triangle, or two triangles. Place an edge of the triangle to the given line, and the straight edge or the other triangle against either of the other edges. The first triangle moved in either direction will present parallel sides.

Perpendiculars and verticals mean two different things. *Perpendicular* is a relative term and means that one line is at 90° to the other, or normal, no matter what the direction of either line. *Vertical*, in a drawing, means a line which is perpendicular to a horizontal one only. In space it is a normal to a horizontal plane, or in other words, to the earth's surface.

Perpendiculars can be made most readily with the triangles and T sq, or straight edge. In the case of oblique lines, place one triangle against the given line and move it against the other triangle, as for parallel lines, until it is a short distance away, then use the second triangle against the first so that the perpendicular drawn can be made to intersect and cross the first, if necessary.

It is a mistake to draw the perpendicular when the first triangle is in contact with the given line.

Right angles can be divided readily into halves and thirds by means of the 45° and 30° and 60° triangles, either with the assistance of a T sq., or without, by the process just described for perpendiculars and verticals. Hence, a circle can be divided into four, eight or twelve parts, and by bisecting one angle a new base can be obtained for halving all the angles and doubling the number of the above mentioned divisions.

Tangency in geometry means identity of direction at a common point. Identity of direction also means that the common tangent to two curves is a normal to the radius of curvature of each at the point of tangency.

Since in geometry lines have no thickness, it follows that two lines in a drawing that are to be tangent to each other should be made not osculating lines, but identical, that is, the thickness of the lines at the point of tangency should be that of the thickest line only which is used.

75. To draw a tangent to an irregular curve at a point on the curve. See Fig. 74.

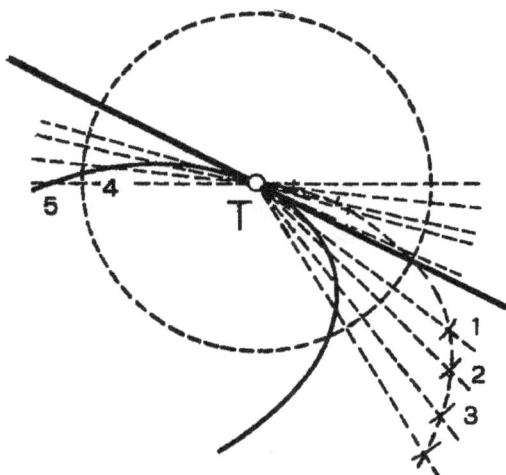

Through T, the point of the desired tangent, draw random secants, as 1, 2, 3, 4, 5, etc., through points on the curve. With T as a center, and any radius, describe an arc to cut the secants prolonged. On each secant lay off, from its intersection with the circle, a distance equal to the chord length of the secant within the irregular curve, and measuring on the same side of the circle as the secant with respect to the point T. Draw a curve through these points. Where this curve cuts the auxiliary circle, is a point in the tangent.

76. To rectify an arc of a curve or of a circle subtending a small angle. See Fig. 75.

Let BA be the given arc. Prolong the chord AB to O, making OA = AB ÷ 2. With radius OB draw an arc to cut a tangent to the curve at A in C. AC will be the

FIGURE NO. 75.

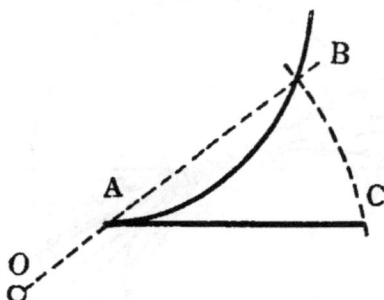

desired length.   This is an approximation, and useful only for short arcs.

77. To draw an arc of given radius tangent to two given oblique lines. See Fig. 76.

FIGURE NO.76.

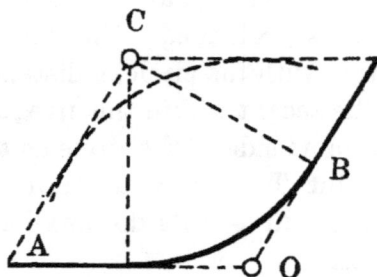

Prolong the given lines to meet at O. With O as a center, and the given radius, describe an arc. Parallels to the given lines, and drawn tangent to this arc, will meet at C, from which perpendiculars to the given lines give the points of tangency, C being the center of the arc.

78. The conics.

The conic sections, or simply conics, appear frequently in mechanics and machine construction. They are called conic sections because they are the contour forms of plane sections of a cone of revolution. A cone of revolution is a cone formed by one line revolving about another, which it intersects, and with which it maintains a constant angle.

FIGURE NO. 77.

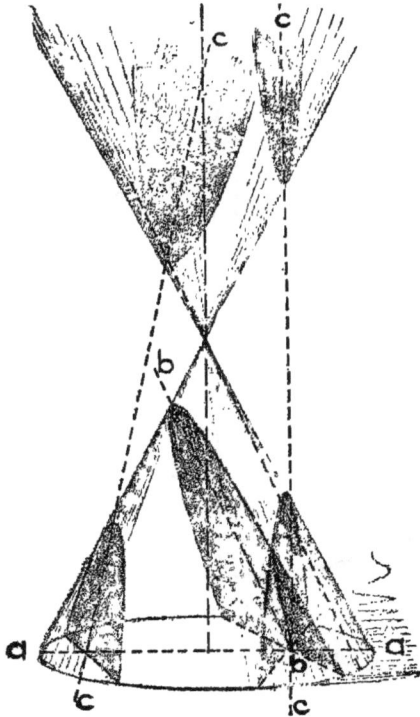

A circle is formed by cutting the cone perpendicular to the axis, as *a—a*, see Fig. 77.

A parabola is formed by cutting the cone in a plane parallel to the moving line as *b-b*.

A hyperbola is formed by cutting the cone in a plane making a less angle with the axis than the moving line, as *c-c*.

An ellipse is formed by cutting the cone at any other angle, i.e., making a greater angle with the axis than the elements.

Ellipses, parabolas and hyperbolas may be drawn in several different ways. A few of the more common and convenient will now be given:

FIGURE NO. 78.

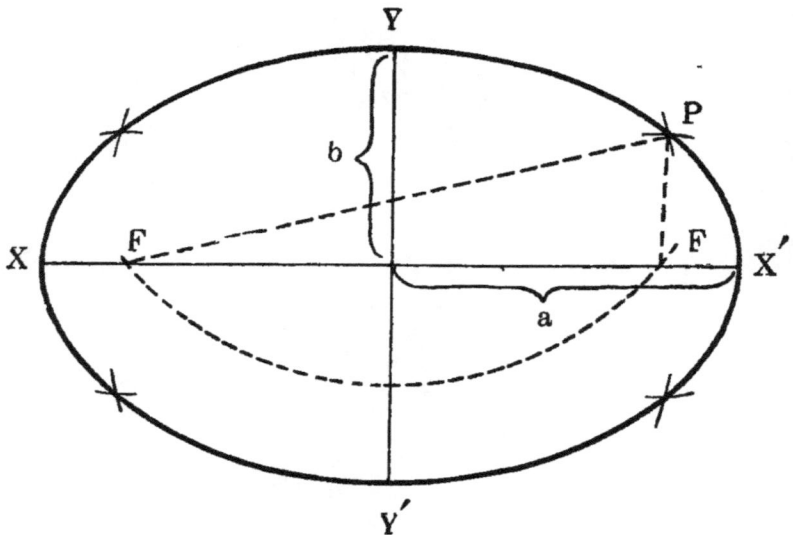

79. To draw an ellipse by the focii method; to draw it by the principle, in other words, that the sum of the focal radii to any point on the curve is a constant, see Fig. 78. The constant is equal to 2*a*. The focal radii are F P and F′ P, respectively.

When the major and minor axes are given, the focii may be obtained as follows; With a radius equal to the semi-major axis, and a center at either extremity of the minor axis, describe arcs to cut the major axis in points which will be the focii, for $FY + YF' = 2a$. To find any point on the curve, as P, take any radius not less than $X'F$ or XF', and with center at F, describe an arc of a circle; with a radius equal to the difference between $2a$, and the radius just taken, and a center at F', describe an arc to intersect the first one at P, which will be a point on the curve. There are three other points on the curve, corresponding with P, and which are symmetrical with respect to the axis, hence, the practical method of procedure is to arbitrarily divide the major axis between the focus and the center of the ellipse into several parts, each to give two radii for four symmetrical points on the curve. With each radius taken, the four symmetrical points are obtained by striking arcs from both focii above and below the major axis.

For careful drawing, more points will have to be plotted in the neighborhood of the major axis, than in that of the minor. It is well to find a group of symmetrical points, as just described, complete, before proceeding with any construction for other points.

80. To construct an ellipse by the method of the trammel.

The trammel is an instrument for mechanically constructing an ellipse, not very successful practically, because of its lack of adaptability, and its cumbersomeness, as well as large cost. It consists, fundamentally, of two tracks or guides at right angles to each other, and which constitue the major and minor axes. A third member, an arm, carries a marking point, while two wheels, or lugs, fastened to it rigidly, move in the grooves in the first two members, and hence constrain the moving arm so that its marking point goes in the path of an ellipse. The operation can be more readily seen after a description of its practical equivalent, see Fig. 79.

FIGURE NO. 79.

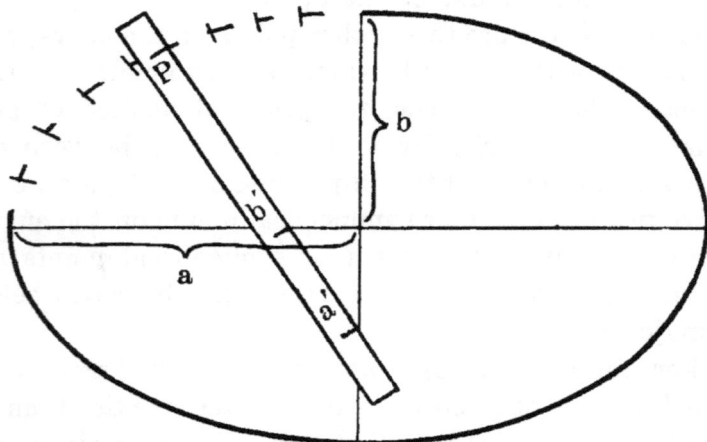

Take a piece of paper as a straight edge, and mark on it a point P. From P lay off a distance, Pa', equal to a, and also a distance, Pb', equal to b. Then with b' touching the major axis, and a' the minor axis, P will be a point upon the curve. To plot points then, move the straight edge around into as many positions as desired, and for each point plotted, indicate its place by a short stroke along the straight edge, and one perpendicular to it at P. This kind of stroke will identify the points most successfully. This method of plotting an ellipse is an excellent one, because of the ease with which the points can be located, where wanted.

81. To draw an ellipse approximately with the compass. See Fig. 80.

FIGURE NO. 80.

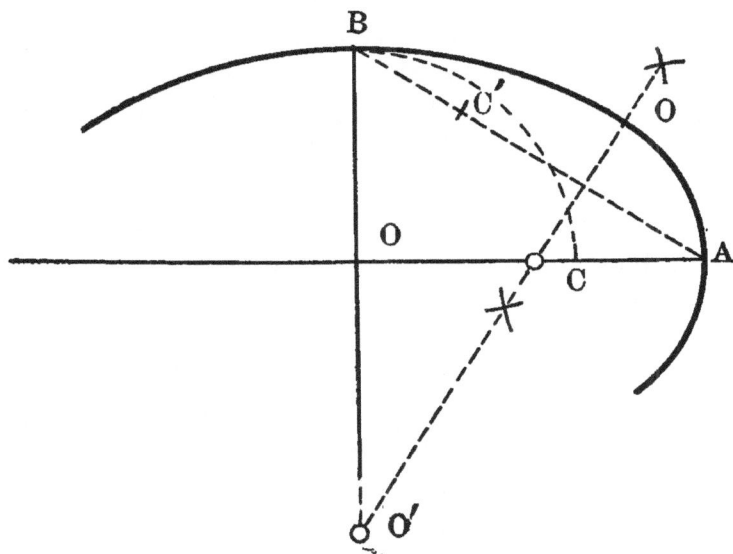

This is the method the mechanical draftsman will alway use, where possible, and it is a very good substitute. There are two ways, the following being the more common:

Draw the minor auxiliary circle, it will cut the major axis in C. Connect B and A. Lay off on BA from B, the distance BC',= AC. Next, bisect the line, C'A, and prolong the bisector, OO', to meet the minor axis prolonged in O'. O' is the center for an arc passing through B, which approximates the ellipse. From the point in which this arc touches the bisector, OO′, draw an arc of another circle, passing through A, which has its center where the bisector cuts the major axis. To complete the curve, another center may be found on the other side of the major axis, and symmetrical with O′, and one on the other side of the minor axis, symmetrical with the center C. This is a method of drawing an ellipse, which, of course more closely simulates the true curve, the nearer the ratio of the two axes is equal to one. If one has any facility in free-hand work, it is recommended that a quadrant be sketched in roughly, and then copied as closely as possible with arcs, and using as many centers as needed for the purpose. The ellipse will not be as true as if plotted by points, but it will be a smooth curve and make a better general appearance than if constructed as first directed.

82. To draw a parabola by means of the focus.

A parabola is defined in mathematics, as the locus of a point which moves, so that its distance from a fixed point, called the focus, is equal to its distance from a fixed line, called the directrix.

Let D D′ (Fig. 81) be the directrix, and O F at right angles to it, the axis. Let F be the focus. Since the distance of F from any point on the curve is equal to the distance of that point from the directrix, to find any point on the curve as P, draw a line parallel to D D′ at any chosen distance from it. Then with this same distance as a radius, and F as a center, describe an arc to cut the parallel in the point, P. Two such points, P and P′, will be found symmetrical with the axis. One point must lie on the axis half way between the focus and directrix, namely, O, which is called the vertex. The entire curve is symmetrical on its axis.

FIGURE NO. 81.

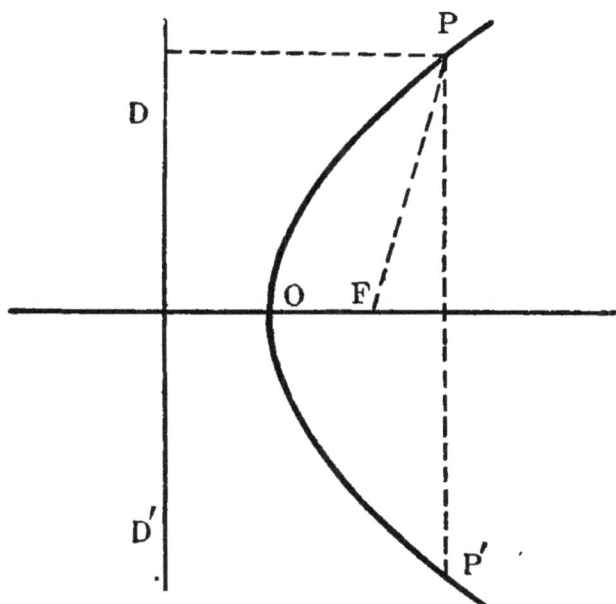

To plot a large number of points, for a good, smooth curve, divide the axis, arbitrarily, into a number of points starting at the vertex. Through these draw parallels to the directrix. Then with radii equal, respectively, to these distances from the directrix, describe arcs from the focus as a center to cut the parallels in points of the curve.

83. To draw a hyperbola by means of its focii.

A hyperbola is defined in mathematics as the locus of a point which moves so that the ratio of its distance from a fixed point, called the focus, **to** its distance from a fixed line, called the directrix, is a constant and greater than unity, see Fig. 82.

FIGURE NO. 82.

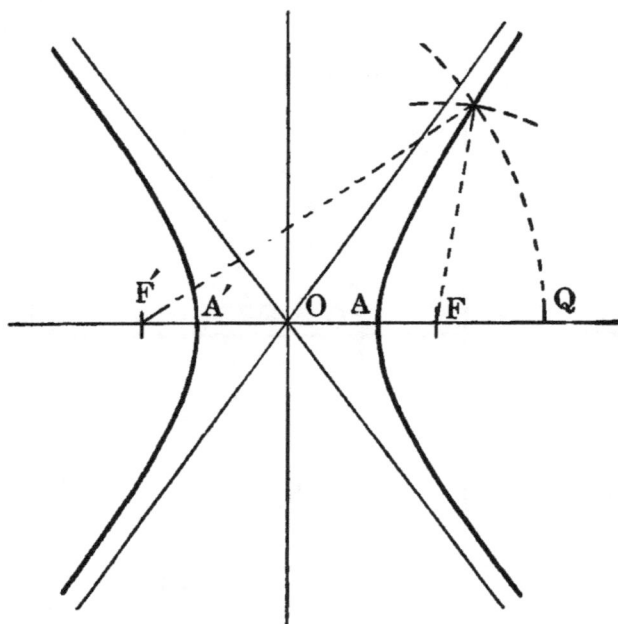

We shall work the problem, however, by the geometrical principle that the difference between the focal radii to any point on the curve is a constant. Let F F′ be the focii, and O the center. Assume the vertices of the curve, A A′, at equal distances from O on the axis F′F. A A′ is also the constant difference. Take any point on the axis a distance from F′ greater than F′A as F′Q, and with F′ as a center describe an arc; then, substracting the constant A′A, from F′Q take the remainder as a radius, and with F as a center describe an arc to cut the first one

in points of the curve above and below the axis, and so
proceed for as many points as desired.

Now, consider that hyperbola, which is formed by a
section of a cone parallel to its axis, it can be easily seen
that if the two elements of the surface (positions of the
moving line), which are parallel to the plane of the section,
were projected on the plane of the section, the curve
would approach but never touch them.   These two
elements are known as the asymptotes of the curve.

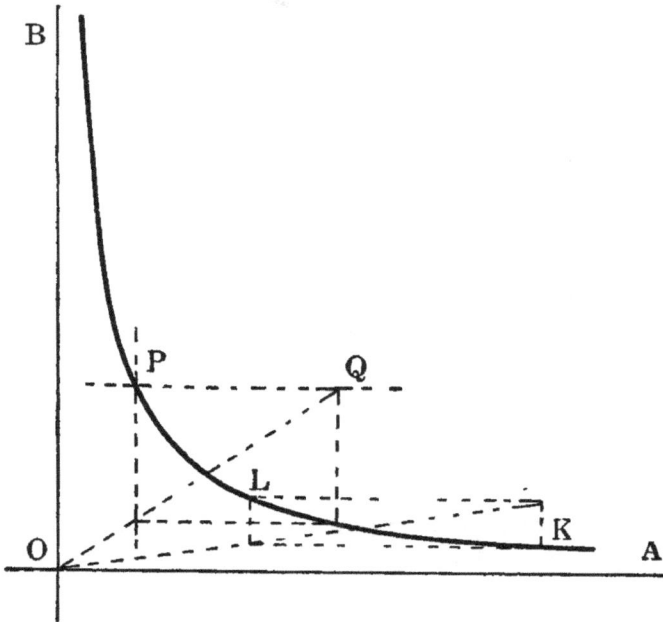

84. To draw a hyperbola by the rectangle method.

This is a method which is associated with certain
properties of steam, when its performance in an engine
cylinder is plotted in a curve, see Fig. 83.

Let AO and OB be two reference lines at right angles to each other, and let P be an assumed point of the curve. Through P draw lines parallel to OA and OB, respectively; also, draw any line, OQ. From the points in which the line, OQ, cuts the parallels through P to OA and OB, respectively, erect perpendiculars, and these will intersect in a point on the curve, and so for as many points desired.

To find the point, O, given any two points, as K and L, on the curve and one of the reference lines: Draw a parallelogram upon K and L as vertices similar to the one erected at P. A diagonal of the parallelogram will cut the given reference lines in the point, O. The latter construction is in common use when applying the hyperbola to indicator cards of engines.

85. The cycloid.

The cycloids belong to a class of curves otherwise known as roulettes. *A roulette* is the path traced by a point upon a curve which rolls upon another curve, the latter being fixed. The rolling curve is the *generatrix*, and that upon which it rolls is called the *directrix*. A *cycloid* may be described as a curve traced by a point upon a circle, which rolls on a straight line or another circle; although the cycloid, proper, is understood to be that curve traced by a point upon a circle which rolls upon a straight line, and the distinguishing terms, epicycloid and hypocycloid, are used to mean the path of a point upon a circle which rolls upon the outside and the inside of another circle, respectively.

The cycloidal curves are those used extensively for the outlines of gear teeth, and every draftsman should at least be familiar in the beginning with their construction. The aim in the construction of gear teeth is to get rolling contact.

86. To construct the cycloid, see Fig. 84.

Let AB be the fixed line or directrix upon which the circle, EP''C, rolls. Suppose the circle to move in the direction of the arrow. When the circumference has

FIGURE NO. 84.

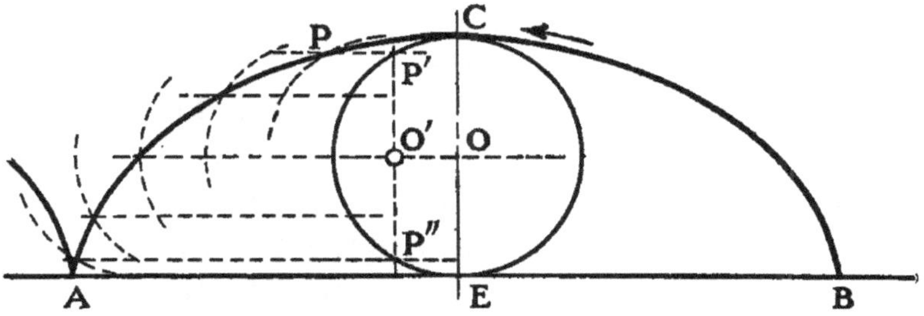

rolled a given fraction of its length upon the line AB, the center, O, will have moved a linear distance equal to this, or to O', that is, OO' = CP' = EP''. The point P'' will lie on the directrix and the point C will have moved the distance towards the directrix that P'' is from it originally. Hence, draw a line parallel to the directrix through P', and with O' as a center and radius equal to the given circle, describe an arc to cut this parallel in the point P, which will be a point of the curve. Repeat this process for as many points as desired.

The curve is symmetrical upon the vertical line, EC. The tangent to the curve at C is parallel to the directrix AB, the tangent at A is perpendicular to the directrix. If the circle continued to roll on AB it would generate another loop, and the point A, would be a cusp of the curve. In gear design, only a small portion of the curve, in the neighborhood of A or B, would be used for the line of the tooth.

87. To draw the epicycloid:

The epicycloid, and also the hypocycloid, are the more common curves for gear teeth, the cycloid being limited to the teeth on a rack. Let A E B, with F as center, Fig. 85, be the directrix, and E D C, with O as center, be the generatix. If the circle E D C rolls through a given circumferential length upon the directrix toward the left, the center O will travel a greater distance, proportional to its distance from the center F. Assume the circumferential length to be E P or one-twelfth of the circumference. P will come down to the directrix at P′ and C will move one-twelfth of the circumferential length to the left of C, or to a distance from the directrix that P′′ is. Hence, through P′ draw a radius of the directrix upon which the new position of the center O will lie, or O′, and at O′ draw an arc of the rolling circle in its new position, also with F as a center, draw an arc of a circle of radius FP′′. Where these two arcs intersect in Q, is one point in the epicycloid. By similar process, as many points can be plotted as desired. The curve is symmetrical upon the line FO. The tangent to the curve at C is normal to the radius F O C, and the tangent at A is normal to the directrix.

88. To draw the hypocycloid, see Fig. 85.

The hypocycloid is similarly constructed to the epicycloid, the generatrix, or rolling circle, moving on the

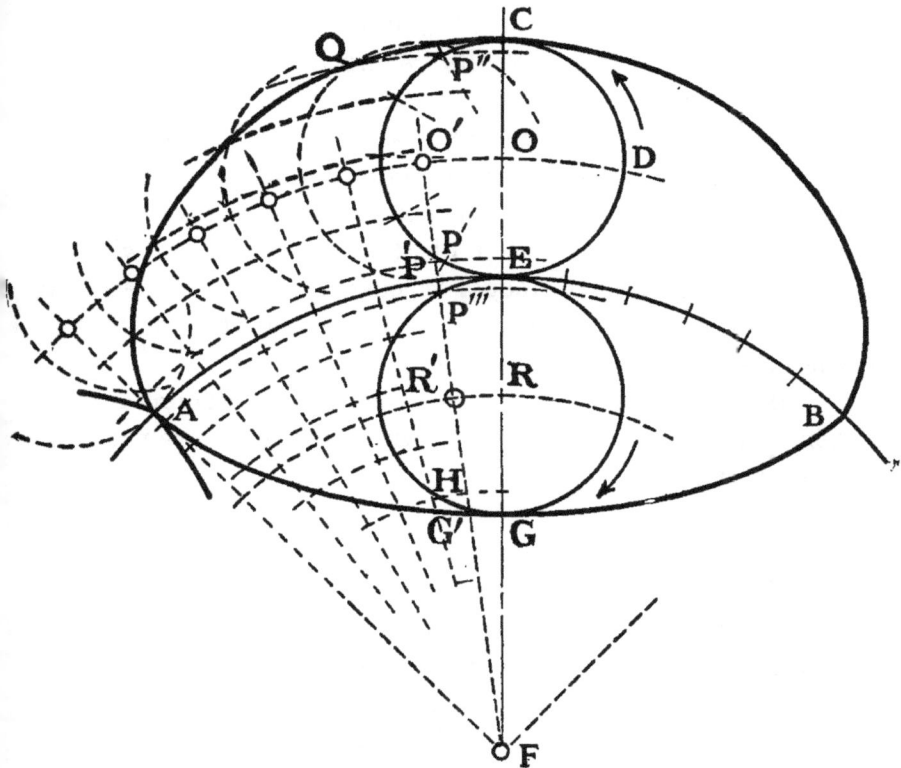

inside of the directrix. The epicycloid is the curve of the
face or upper half, the hypocycloid, the flank or lower
half of the tooth outline. The center R of the generating
circle GE., it will be seen, travels through a shorter
linear distance than the points on the circumference.

When the circumference has rolled off a distance, for
example EP''', the center will have traveled to the position
R' and the point G will have traveled one-twelfth of the
circumferential length to the left or to the distance from

the directrix that H is.   Hence, through P′′′ draw a radius
of the directrix upon which the new position of the center
of the generating circle will lie, and with this point, R′, as
center, describe an arc of the generating circle; also with
F as a center describe an arc of a circle of radius FH.
Where these two curves intersect, is a point in the
hypocycloid, and so on for as many points as desired.   If
the circumference of the generating circle will go an even
number of times into that of the directrix, there will be
that even number of loops, or cusps.   If the generatrix
does not go an even number of times into the directrix,
then the cusps will not close entirely.

Both epicycloid and hypocycloid have common tangents at the points, A and B, which tangents are radii of
the directrix.

89.   To draw the involute of a circle.

The involute of a circle may be defined as the path
traced by any point of its circumference if the latter is
conceived to unroll from the circle as a thread from a
spool.   Or again, geometrically, it is the locus of the
extremities of a series of tangents to the circle, starting at
some fixed reference point and whose lengths are respectively equal to the circumferential lengths between the
points of tangency and the reference point.   It is a curve,
which like the cycloids, finds illustration in forms of gear
teeth, a gear which works on a rack.

Let A B C, Fig. 86, be any circle with A a fixed
reference point.   Divide the circle conveniently into a
number of equal divisions.   Starting at the first division,
on either side of the reference point, draw a tangent and

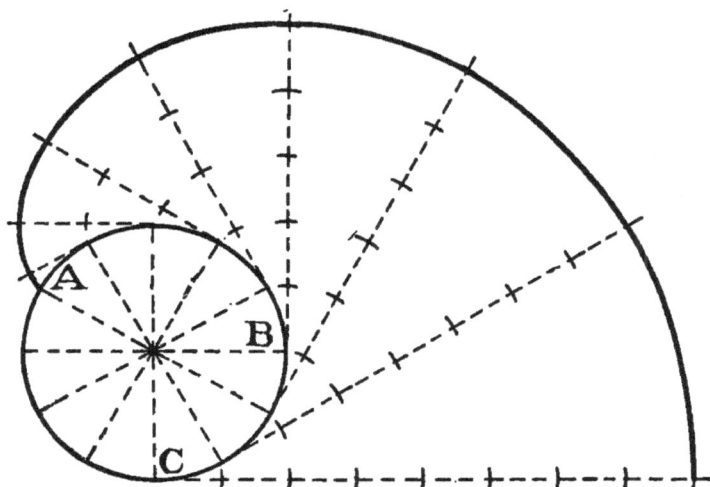

make its length equal to the length of the arc between the point of tangency and the point A. Its extremity is one point on the curve of the involute. Next, draw a tangent at the second equal division and make its length equal to two segments of the circumference, establishing by its extremity another point in the curve, and so on for as many points as desired necessary to draw a smooth curve through them.

90. Geometrical definitions, terms, etc.

Curves, in a mathematical sense, include straight lines as well as curves. A straight line is a curve of infinite radius.

A plane figure is a plane bounded on all sides by lines. If the lines are straight, the space which they enclose is called a rectilinear figure, or polygon, and the sum of the bounding lines is the *perimeter* of the polygon.

Polygons are named according to the number of their sides as a triangle, quadrilateral, pentagon, hexagon; a heptagon, of seven sides; octagon, of eight; nonagon, or enneagon, of nine; decagon, of ten; undecagon, of eleven; dodecagon, of twelve. Polygons are supposed to be regular unless otherwise stated.

A diameter of a polygon is any line drawn through the center of a figure, and terminated by the opposite boundaries.

The long diameter of a polygon is the diameter of its circumscribed circle. This is also called a diagonal.

The short diameter of a polygon is the diameter of its inscribed circle.

A polyhedron is a solid bounded entirely by planes.

There are only five regular polyhedrons. viz.:

The tetrahedron, bounded by four equilateral triangles.

The hexahedron, or cube, bounded by six equal squares.

The octahedron, bounded by eight equilateral triangles.

The dodecahedron, bounded by twelve equal pentagons.

The icosahedron, bounded by twenty equal equilateral triangles.

A prism is a polyhedron having two of its faces, called its ends or bases, parallel, and the rest parallelograms.

A parallelopiped is a prism whose bases are parallelograms.

The axis of a prism is a straight line joining the centers of its bases.

The axis of a pyramid is a straight line from its vertex to the center of its base.

A right prism, or pyramid, has its axis at right angles to its base.

CHAPTER VII.

MACHINE SKETCHING.

91. Machine sketching.

One of the most important accomplishments an engineer
can have is the ability to make a free-hand working
drawing sketch. The machine designer finds it invaluable
if he can turn his sketches over to the junior draftsman to
work up. The junior draftman finds it valuable if a part
of a machine has to be repaired, and temporary drawings
made of it, or again, to make a record of construction for
future reference.

No particular scale is in general required in sketching,
but a proper proportioning of each view and its size
relative to other views.

To get the proper relation of values, i.e., distances and
sizes of lines, a certain ratio must be obtained. The ratio
in perspective drawing or picture making is a very different
problem from that in mechanical drawing. In the former
the ratios are between sizes that are not of certain true
length, but apparent lengths due to the obliquity of these
lines to the observer, or foreshortening, as it is called,
while in mechanical drawing these are absolute sizes being
the practical lengths, as a rule, that are in the subject.—

Therefore, to obtain a value to record in drawing that
is a ratio between two quantities, estimate the value in the
original by assigning either a numerical ratio of simple
numbers easily transferred, or else, carry a conception

of the ratio value in the mind without assigning any numerical value. For illustration, if a certain detail is one-fifth as long as the over all dimension, then that detail in the drawing should be one-fifth as long as the overall. In some cases of rather large subjects it may be advisable to make a space measurement of sizes for comparison. For illustration, if a certain detail is one-fifth the length of an over all, a pencil may be held up between the eye and subject, and covering one end of the detail to be measured with the end of the pencil, and the other end by the finger moved along the pencil, this length may be compared with the over all by noting that it will go into it five times. Therefore, as the pencil unit is to the whole unit stepped off with it, so the detail in the drawing should be to the over all in the drawing. This method can be used where the quantities to be measured do not extend throughout a very wide range of vision, else the foreshortening, before mentioned, will cut a very serious figure and introduce inaccuracies. An eye estimate of values is better than an actual measurement made. After the drawing has all been made, however, its accuracy may be tested by these means very nicely.

Sometimes sketching may take the form of a hasty drawing or drawings of a model from which afterwards working drawings can be made. It is possible, generally, to get the data into two or three views. The problem is to choose the fewest views, and some discrimination has to be exercised to ascertain what they are. Or it may be that the final drawing is to be one which shows all the constructive facts —without being a working drawing—with the fewest views. Such a problem is illustrated by a patent office drawing.

Sketching may be done upon regular sheets, but more frequently it is put upon scratch paper, or the pages of a note book. Use of a rule and compas is here not feasible. If done on the pages of a note book, the first problem is to adopt such a size that the subject may be treated properly and legibly. It may involve only one view on a page, or it may involve two or more. If there is room for more than one view on a page they should be sketched in the same projective relation to one another that they would be if mechanically drawn, and upon the different sheets of the note book, as far as possible, the views should be sketched in the same position that they would occupy, if drawn mechanically on a sheet; otherwise, a little confusion in reading the sketches may result. Where several views are to go on one page, in the sketching they should be roughly blocked out for proper allotment of space before doing further work upon any one view; where feasible, sketch them together throughout.

In very rare cases a whole view even cannot be put on a sheet. It should be stopped at a conventional line, such as was described in the case of sections where the plane of the section moved from one place to another, and it is furthermore best to sketch the two or more parts on the sheet in the same relative position so that, if called for, they may be cut along the line of their separation and fastened together to make the complete view.

Views of parts should be sketched as occupying the relative position which they would in the machine.

Some further practical points about sketching follow:—

(*a.*) Throughout sketching, observe, carefully, right angles and the rounded corners; the latter may, in the preliminary operations, be made sharp, and only in the final stage rounded off to the desired amount. Be sure that the right angles are always as near right angles as they can be made.

(*b.*) Where forms are symmetrical upon a center or center line, sketch the two or more parts at the same time. The brief preliminary record made upon one side should be followed by the same record on the other side, and as each stage is gone through with on the one, it should be followed by the same stage on the other. The mistake is quite common to draw one side of a thing that is symmetrical on an axis and then copy it faithfully on the other.

(*c.*) *Sketches of a subject, if made from a model* or the original, should be finished complete before any dimensions are put on, for the problem of sketching does not involve scale or absolute size; it is better if these are not considered. And it is to be observed that the lines of the sketches should be much sharper and blacker if they are to be dimensioned, than if left without, for dimension lines and figures cover up form. When sketches are ready for dimensions these should be taken from the subject in the way heretofore described for mechanical drawings, noting in particular to take only one dimension at a time.

(*d.*) *Sometimes sketches must be made hurriedly*, when it is very important to subdivide the time to be used into parts to be given each to certain stages of the drawing. Roughly speaking, about as much time should be allowed for dimensioning as that for making sketches. Again, the time alloted for sketching may be divided in parts, three of which to be given to the light perliminary work, and one part to finishing up in clear line, for the proportioning and the shape are more important than the character of the line.

(e.) *If time has been improperly allotted,* or a great deal has to be accomplished in a very short time, it may be desirable to cut steps short by leaving out certain things. Where parts are duplicated, for example, they need not be drawn but once. Where forms are encountered which are perfectly understood, like wheels, only a small portion needs to be sketched. Much can frequently be omitted by giving a few written directions explanatory of them; this is true of holes for bolts, fastenings of a similar character, etc. *Where forms are symmetrical upon a center* line only one-half need be sketched; where symmetrical upon a center perhaps only one-quarter, a note being used to refer to the ommitted parts.

(f.) *A complicated subject may have to be drawn in a very compact space,* and in short time. There may not be room upon the pages of the note book to develop the forms, either entire or in separate parts, and even proper proportioning is not feasible. It may be necessary to sketch neglecting proportions entirely, or it may be convenient to change one dimension without changing the other. Again, it may be that some parts can be shown in proportion, while others are not, compressing the drawing where needed to make it fit the space. These are practical problems that can easily be solved when they arise. With a knowledge of the principles of sketching, it becomes merely the question of recognizing the limiting conditions of the particular problem in hand. The draftsman who expects to do comprehensive work cannot afford to be without a knowledge of sketching.

APPENDIX.

BLUE PRINT PROCESS AND REPRODUCTION.

A mechanical drawing made to be worked from is generally reproduced by blue print, or analogous process. A drawing may be reproduced in photo engraving to furnish a cut for a catalogue.

The following upon the subject of blue prints is taken from F. N. Willson's 'Theoretical and Practical Graphics:'

"A sheet of paper may be sensitized to the action of light by coating its surface with a solution of red prussiate of potash (ferrocyanide of potassium) and a ferric salt. The chemical action of light upon this is the production of a ferrous salt from the ferric compound; this combines with the ferrocyanide to produce the final blue undertone of the sheet; while the portions of the paper, from which the light was intercepted by the inked lines, becomes white after immersion in water."

"The proportions in which the chemicals are to be mixed are, apparently, a matter of indifference, so great is the disparity between the receipts of different writers."

"The entire process, while exceedingly simple in theory, varies, as to its result, with the experience and judgment of the manipulator. To his choice the decision is left between the following standard recipes for preparing the sensitizing solution. The "parts" given are all by weight. In every case the potash should be pulverized, to facilitate its dissolving."

No. 1.   FROM LE GENIE CIVIL.

SOLUTION.

No. 1.   Red Prussiate of Potash_____   8 parts
       Water_____ 70 parts
No. 2.   Citric of Iron and Ammonia_____ 10 parts
       Water_____ 70 parts

Filter the solutions separately, mix equal quantities and then filter again.

No. 1.   FROM THE U. S. LABORATORY, WILLETT'S POINT.

SOLUTION.

No. 1.   Double Citrate of Iron and Ammonia, 1 ounce
       Water_____ 4 ounces
No. 2.   Red Prussiate of Potassium_____ 1 ounce
       Water_____ 4 ounces

"The solutions should be dissolved separately, as then they are not sensitive to the action of light. They should be mixed and applied only in the dark room."

"The best American practice is to apply the solution with a flat brush, and to obtain an even coat by stroking first one way, then at right angles. If necessary, a coat of diagonal strokes may be given to secure evenness."

"To copy a drawing, it is placed in a blue print frame made for the purpose, the sensitized paper, with the sensitized surface outermost and immediately back of the drawing. Exposure for about five minutes to the rays of the sun is usually sufficient to get good results, after which the paper is taken out and placed in a bath of water when the superfluous chemicals are washed off."

"White lines can be drawn upon a blue print by using a solution of soda, potash, quick lime, or any alkali with water, adding a little gum arabic to keep the liquid from spreading. It can be applied with the writing pen, ruling pen or brush, according to the area desired to be white."

INDEX

184   INDEX

.

www.ingramcontent.com/pod-product-compliance
Lightning Source LLC
Chambersburg PA
CBHW021711210326
41599CB00013B/1610